Advances in Artificial and Biological Membranes

Advances in Artificial and Biological Membranes

Mechanisms of Ionic Sensitivity, Ion-Sensor Designs and Applications for Ions Measurement

Editors

Andrzej Lewenstam
Krzysztof Dołowy

MDPI • Basel • Beijing • Wuhan • Barcelona • Belgrade • Manchester • Tokyo • Cluj • Tianjin

Editors
Andrzej Lewenstam
AGH University of Science and Technology
Poland

Krzysztof Dołowy
Warsaw University of Life Sciences – SGGW
Poland

Editorial Office
MDPI
St. Alban-Anlage 66
4052 Basel, Switzerland

This is a reprint of articles from the Special Issue published online in the open access journal *Membranes* (ISSN 2077-0375) (available at: https://www.mdpi.com/journal/membranes/special_issues/Advances_Ion_Sensors).

For citation purposes, cite each article independently as indicated on the article page online and as indicated below:

LastName, A.A.; LastName, B.B.; LastName, C.C. Article Title. *Journal Name* **Year**, *Volume Number*, Page Range.

ISBN 978-3-0365-1070-5 (Hbk)
ISBN 978-3-0365-1071-2 (PDF)

© 2021 by the authors. Articles in this book are Open Access and distributed under the Creative Commons Attribution (CC BY) license, which allows users to download, copy and build upon published articles, as long as the author and publisher are properly credited, which ensures maximum dissemination and a wider impact of our publications.

The book as a whole is distributed by MDPI under the terms and conditions of the Creative Commons license CC BY-NC-ND.

Contents

About the Editors vii

Andrzej Lewenstam and Krzysztof Dołowy
Special Issue "Advances in Artificial and Biological Membranes: Mechanisms of Ionic Sensitivity, Ion-Sensor Designs, and Applications for Ion Measurement"
Reprinted from: *Membranes* 2020, 10, 427, doi:10.3390/membranes10120427 1

Peter Lingenfelter, Bartosz Bartoszewicz, Jan Migdalski, Tomasz Sokalski, Mirosław M. Bućko, Robert Filipek and Andrzej Lewenstam
Reference Electrodes with Polymer-Based Membranes—Comprehensive Performance Characteristics
Reprinted from: *Membranes* 2019, 9, 161, doi:10.3390/membranes9120161 7

Krzysztof Maksymiuk, Emilia Stelmach and Agata Michalska
Unintended Changes of Ion-Selective Membranes Composition—Origin and Effect on Analytical Performance
Reprinted from: *Membranes* 2020, 10, 266, doi:10.3390/membranes10100266 29

Nikola Lenar, Beata Paczosa-Bator and Robert Piech
Optimization of Ruthenium Dioxide Solid Contact in Ion-Selective Electrodes
Reprinted from: *Membranes* 2020, 10, 182, doi:10.3390/membranes10080182 43

Ernő Lindner, Marcin Guzinski, Bradford Pendley and Edward Chaum
Plasticized PVC Membrane Modified Electrodes: Voltammetry of Highly Hydrophobic Compounds
Reprinted from: *Membranes* 2020, 10, 202, doi:10.3390/membranes10090202 61

Miroslaw Zajac, Andrzej Lewenstam, Piotr Bednarczyk and Krzysztof Dolowy
Measurement of Multi Ion Transport through Human Bronchial Epithelial Cell Line Provides an Insight into the Mechanism of Defective Water Transport in Cystic Fibrosis
Reprinted from: *Membranes* 2020, 10, 43, doi:10.3390/membranes10030043 77

Jerzy J. Jasielec, Robert Filipek, Krzysztof Dołowy and Andrzej Lewenstam
Precipitation of Inorganic Salts in Mitochondrial Matrix
Reprinted from: *Membranes* 2020, 10, 81, doi:10.3390/membranes10050081 91

Yan Lyu, Shiyu Gan, Yu Bao, Lijie Zhong, Jianan Xu, Wei Wang, Zhenbang Liu, Yingming Ma, Guifu Yang and Li Niu
Solid-Contact Ion-Selective Electrodes: Response Mechanisms, Transducer Materials and Wearable Sensors
Reprinted from: *Membranes* 2020, 10, 128, doi:10.3390/membranes10060128 121

About the Editors

Andrzej Lewenstam received a Ph.D. (1977) and D.Sc. (1987) from the University of Warsaw, Poland. Since 1990, he has been a Professor of Sensor Technology at Åbo Akademi University, Turku, Finland, and the Director of the Center for Process Analytical Chemistry and Sensor Technology 'ProSens' at this university. Concurrently, he is a Professor in Chemistry at AGH University of Science and Technology, Cracow, Poland. He is also an Affiliate Professor in Chemistry, University of Warsaw, Poland, and an Adjunct Professor in Electroanalytical Chemistry at Aalto University, Helsinki, Finland. He was the R&D Manager and Head of Sensors Development at KONE Instruments, Finland, and was a Senior Scientific Adviser at Thermo Fisher Scientific, USA (1984–2016). He was the Chairman of the Working Group on Selective Electrodes and Biosensors of the International Federation of Clinical Chemistry (IFCC), 2006–2010, and a long-term expert of the IFCC. His major fields of interest include chemical and biosensors, clinical and analytical chemistry, electrochemistry, and mathematical modeling. His separate activity concerns the methodology of chemistry and philosophy of science. He is the author of approximately 250 peer-reviewed papers in chemistry (h-index=53), 30 articles in the philosophy of science, and 25 patents; he is also the co-editor of several books and an editor and a member of the editorial board for several international journals.

Krzysztof Dołowy received a Ph.D. (1976) in chemistry from the University of Warsaw, a D.Sc. (1986) in biology from Nencki Institute of Experimental Biology of Polish Academy of Science, and a state professorship in 1998. Since 1990, he has been a Chairman of Physics Department of Warsaw University of Life Sciences—SGGW. He was a post-doc in Boston, Mass., USA (1977-1978); Glasgow University, UK (1979); Vanderbilt University Nashville, Tenn., USA; and visiting professor in KONE Corp. Espoo, Finland. His major interests include the electrochemical structure of the cell membrane, the properties of ion channels, cystic fibrosis, and ion and water transport across the epithelium. In 1990s, he was a politician and Member of Parliament. His separate activity is popularizing science.

Editorial

Special Issue "Advances in Artificial and Biological Membranes: Mechanisms of Ionic Sensitivity, Ion-Sensor Designs, and Applications for Ion Measurement"

Andrzej Lewenstam [1,2,3,*] and Krzysztof Dołowy [4]

1. Faculty of Materials Science and Ceramics, AGH University of Science and Technology, 30-059 Krakow, Poland
2. Faculty of Chemistry, Biological and Chemical Research Centre, University of Warsaw, 02-089 Warsaw, Poland
3. Faculty of Science and Engineering, Johan Gadolin Process Chemistry Centre, Centre for Process Analytical Chemistry and Sensor Technology (ProSens), Åbo Akademi University, 20500 Turku, Finland
4. Department of Biophysics, Warsaw University of Life Sciences—SGGW, 02-776 Warsaw, Poland; krzysztof_dolowy@sggw.pl
* Correspondence: andrzej.lewenstam@gmail.com

Received: 3 December 2020; Accepted: 10 December 2020; Published: 15 December 2020

Ion sensors, conventionally known as ion-selective membrane electrodes, were devised 100 years ago with the invention of a pH electrode with a glass membrane (in 1906 Cremer, in 1909 Haber and Klemensiewicz). The electrode type had a symmetric design with an internal contact made with a solution. Interestingly, in the same period, membrane biophysics made a great step forward. It was hypothesized that the nerve cell is covered by an invisible membrane specifically selective to potassium ions. Since the concentration of potassium ions in the cytoplasm is high, and in the external medium low, the membrane potential is negative and during nerve impulse it nullifies (Bernstein, 1912).

Today, in ion-sensor technology, there are many membranes made of glass, solid-state, plastics, and composites, as well numerous internal contacts made of conducting polymers, carbon nanotubes, graphene, conducting clays, and composites. Owing to these advances, sensors can be miniaturized, made service-free, and produced by a mass fabrication technique, such as 3D printing. The sensor responses are now interpreted by models that are able to access the time and space domains. In this way, sensor signals may be mathematically transposed. Supported by advanced modeling, ion sensors are now deliberately calibration-free, and quality checking can be automated. The sensors can be used in ad hoc and routine application areas like clinical care and diagnostics, monitoring personal physiological performance in sport, and measuring water quality.

The heart of the ion sensors is always the same: a membrane that can develop the ion response due to selective, fast, and reversible ion exchange at the sample–membrane interface. The membrane can be artificial or biological.

This Special Issue presents major progress that has been made recently in ion-sensor technology. Scientific activity in this field is currently very intense and is characterized by several hundred published papers. The editors have by no means aimed to cover exhaustively all subjects. We, the editors, briefly elucidate the background to the recent trends in order to introduce the reader to the more detailed reports by different authors with references, which are gathered here in this Special Issue volume.

Conventional membrane ion-selective electrodes (ISEs) contain the ion-sensitive membrane that separates the internal contact (with connecting wire to the electrometer) and solution (containing the analyte of interest). From the very beginning, the electrodes with a solid membrane containing insoluble inorganic salts were equipped with a solid contact (metal wire, e.g., Ag or Pt, carbon rod) while the ISEs with liquid and derived plastic membranes containing ion exchangers or neutral ligands were equipped with a so-called internal solution and internal reference electrode, usually Ag, AgCl electrode, immersed in the solution. Interestingly, historically the first ISE pH electrode employed a glass membrane which is an overcooled liquid in solid-state at room temperature. The internal solution typical for this popular electrode ignited a search for a solid contact by the Nikolskii group, known for their pioneering contributions dedicated to the glass electrode. The idea was to find a material that could couple hydrogen ion sensitivity. For this purpose, alkaline metal alloys (e.g., Li or Na in Sn) were used to coat the inner side of the glass membrane used [1].

The elimination of the internal solution in ion-selective membrane electrodes and substituting with plastic membranes (usually poly (vinyl chloride), PVC) and in reference electrodes has proved by far to be challenging. This approach aroused intense research interest, which has lasted up to today. No wonder, since the ISE plastic membranes are known for their sensitivity to ions of physiological importance, like sodium, potassium, calcium, magnesium, chlorides, and bicarbonates, which are extensively measured in medical theatres. The need for miniaturization of these ISEs while keeping them robust, durable, cheap, and maintenance-free accompanied the expectation of automated clinical measurement. The answer to this challenge was offered by substituting the liquid contact with a solid contact made from a conducting polymer (CP) layer, as reported by Lewenstam and co-workers in 1992 [2]. The first solid-contact sodium-selective electrode with a plastic membrane was described, thus opening a new chapter in the electrochemical analysis of ions, especially blood electrolytes. Conducting polymer is a versatile material that addresses several demands in ion-sensor technology. CP is a solid phase and does not evaporate. CP is characterized by mixed conductivity, intrinsic ion-to-electron coupling, it is a material with free electrons mediating at the CP-lead to meter interface and offers the possibility of intentional ligand doping which allows tailoring the desired ionic sensitivity by ion exchange with bathing solution. Additionally, CPs can be soluble and processable.

On its own, a CP membrane can be an electrode membrane for anions, cations, or electrons (redox electrode) [3,4] or an ionically insensitive CP-based membrane reference electrode [5]. CP can also play a role of ion-to-electron transducer, when covered with an ISE plastic membrane [6,7]. In recent years, the concept of a solid contact is reinforced by use of nanostructures deposited in thin layers of, e.g., carbon in different 3D forms and of carbon doped by, or directly used, nanostructures of metals, transition (redox) metal oxides, salts and ligands with redox groups, or carbon nanostructures with CPs. In this way the electrochemical charge coupling could be downscaled to nano. The materials, their benefits and limits are known from supercapacitors. Developing an electrochemically accessible surface area and downscaling to nano dimensions increase the interfacial capacitance of the solid contact, and the apparent stability of the contact is observed [8,9]. A new wave that just enhanced solid contact technology is represented by the use of 3D printable composite PVC-based materials as the electrochemical membranes of all-solid-state chemical ion-sensors and reference electrodes [10]. Since PVC ISE membranes and reference composites contain hidden water, the sandwich of these two elements forms a "well-wetted" solid contact [11].

As mentioned, the papers collected here reflect the main trends and challenges, actual applications and potential.

In the very empirical report on plastic membranes with dispersed equitransferent electrolytes, Lingenfelter and co-authors [12] showed how to obtain the reference electrodes. The research in this area is governed by the idea that there is a wide insensitivity towards ions in the matrices, which is opposite to the aim of ion sensors. Since the reference electrode in experimental applications typically serves several indicator ISEs, the research in ISEs is distinctively more laborious in comparison to indicator electrodes. The report shows for the first time in detail a comprehensive, semi-industrial multi-solution protocol on testing the reference electrodes and the criteria for selecting the most suitable one.

There are three papers that represent crucial pre-set topics in ion sensors.

Maksymiuk et al. [13] review the unfavorable effects of dissolution of the solid-contact material in the plastic membrane. In a convincing way, the authors underline a possible mutual influence of CPs on the properties of plastic ISE membrane and vice-versa. Lenar et al. [14] describe an application of ruthenium oxide as the mediating material in a solid-contact potassium-sensitive ISE. The authors point out the increase of redox capacity and response reproducibility when the hydrated ruthenium oxide is used.

There is an intriguing link between these two papers which directly inspires a new research pathway. For a long time, there has been the belief that an ideal solid contact should not allow a water layer formation between the solid contact and the ISE membrane (even if it is known that PVC ISE membranes contain "hidden water"). A strong theoretical argument is supported by so-called water tests which show the sluggish response if the water layer is present at the contact–membrane interface. Maksymiuk et al. [13] correctly argue though that the conventional electrodes with internal solution are by far more stable: "a classic ISM-containing arrangement offers high stability of potential readings in time due to well-defined, reversible ion-electron transfer through all sensor interfaces". On the other hand, in the paper of Lenar et al. [14], we can read that "the occurrence of structural water in $RuO_2 \bullet xH_2O$, which creates a large inner surface available for ion transport and was shown to be a favourable feature in the context of designing potentiometric sensors". Accordingly, the authors provide evidence that the solid contact with chemically bounded hydration water in it can be as good as highly hydrophobic. The former case does not allow for the water layer formation, and the latter has sufficient water content to stand water transport through the membrane decreased by a small gradient of chemical potential for water between the "well-wetted" solid contact and ISE membrane (the same argument can be applied for ion transmembrane fluxes and mechanical membrane–contact adhesion making a water test redundant). In this situation, as correctly commented by Maksymiuk et al. [13] an optimization of the sensor with solid contact depends on a lot of factors and in the last instance the experimental theatre in which the ion-sensors are used.

Lindner et al. [15] present the application of ion-sensitive membranes in nonzero-current measurement. The authors convincingly demonstrate the utility of the plasticized PVC membrane modified working electrode for the voltammetry measurement of highly lipophilic molecules. Nonzero-current measurement has an additional value vs. the zero current mode described above, since the ISE response can be described without ambiguity by equivalent circuits. In-zero current, i.e., potentiometric ISEs, the equivalent circuit of a pseudocapacitor, implying faradaic processes at solid contact–membrane interface is predominantly used. However, recently, the more rigorous double-layer capacitor, with its simple equivalent circuit reflecting charge interactions only, is as well considered.

Finally, in the review by Lyu et al. [16] the responses of solid-contact electrodes are highlighted by listing the various materials used and providing a very basic insight into theoretical principles. The authors dedicate their preview to a hot issue of present sensor technology, namely wearable sensors (see below).

Wearable solid-contact sensors, their fabrication, challenges, and perspectives are reviewed by Lyu et al. [16]. The authors show several formulations that can serve in the screening of electrolytes in sweat. The development and market are highly anticipated due to growing popularity of sports like the marathon or triathlon. Monitoring body dehydration or electrolyte imbalance plays a crucial role in the well-being and success of participants.

ISEs are often used for automated determination of ion concentration in whole blood or plasma. In such measurements, the electrode is usually washed and calibrated before each measurement within a few seconds. Zajac et al. [17] show that the use of ISEs might be a reasonable choice to study ion transport across epithelial cells. However, the number of ions transported is very small. The only possibility to measure the change in ion concentration by an ISE in the solution facing the living cells is to drastically reduce its volume [18]. In addition, the washing and calibration procedure for an ISE facing living cells must be changed. The ISE must be calibrated before the experiment and when mounted on the measuring platform the one-point calibration in the cell's friendly medium can be performed. The ISE used must be miniaturized and be able to survive for an hour of measurements without its properties changing. The use of sodium, potassium, pH, and chloride electrodes provided insight into the mechanism of ion and water transport across the epithelium.

The Jasielec et al. [19] paper is a theoretical one that describes the transport of ions into living mitochondria. Since mitochondria have a high negative inner potential caused by expulsing proton ions at the expense of metabolic energy, they are prone to inward transport of cations into the matrix. The transport of cations into the matrix occurs via ion channels present in the mitochondrial membrane, which are precisely controlled by physiology. Mitochondria are especially effective in transporting calcium ions into the matrix. Since at the same time phosphates are also transported, calcium phosphate gel is formed in the matrix. The paper of Jasielec et al. [19] analyze the role that magnesium, carbonate, and hydroxyl ions play in the process of formation of osmotically inactive gel in the matrix formed in the multi-ion complex environment.

It would be rewarding to penetrate a mitochondrial membrane to measure free magnesium and calcium in the matrix. This calls for nanoelectrodes. Are they beyond limits?

Without a doubt, the electroanalytical application of membranes has come to a new period which is marked by the advent of the anisotropic heterogenous composite membrane. Therefore, this breakthrough brings with it the need for future research in 3D modeling, 3D print-fabrication, multiscale applications, and associated multidimensional challenges.

Funding: This research received no external funding.

Acknowledgments: The editors would like to thank the authors for their high-quality outputs and Angela Yang for her patient and motivating help and assistance.

Conflicts of Interest: The authors declare no conflict of interest.

References

1. Nikolskii, B.P.; Materova, E.A. Solid contact in membrane ion-selective electrodes. *Ion Select. Electr. Rev.* **1985**, *7*, 3–39.
2. Cadogan, A.; Gao, Z.; Lewenstam, A.; Ivaska, A.; Diamond, D. All-solid-state sodium-selective electrode based on a calixarene ionophore in a poly(vinyl chloride) membrane with a polypyrrole solid contact. *Anal. Chem.* **1992**, *64*, 2496–2501. [CrossRef]
3. Lewenstam, A.; Bobacka, J.; Ivaska, A. Mechanism of ionic and redox sensitivity of p-type conducting polymers. *J. Electroanal. Chem.* **1994**, *368*, 23–31. [CrossRef]
4. Migdalski, J.; Blaz, T.; Lewenstam, A. Conducting polymer-based ion-selective electrodes. *Anal. Chim. Acta* **1996**, *322*, 141–149. [CrossRef]
5. Blaz, T.; Migdalski, J.; Lewenstam, A. Junction-less reference electrode for potentiometric measurements obtained by buffering pH in a conducting polymer matrix. *Analyst* **2005**, *130*, 637–643. [CrossRef]
6. Huang, M.R.; Guo-Li, G.U.; Ding, Y.B.; Fu, X.T.; Li, R.G. Advanced solid-contact ion selective electrode based on electrically conducting polymers. *Chin. J. Anal. Chem.* **2012**, *40*, 1454–1460. [CrossRef]
7. Bobacka, J.; Ivaska, A.; Lewenstam, A. potentiometric ion sensors. *Chem. Rev.* **2008**, *108*, 329–351. [CrossRef]
8. Mousavi, Z.; Teter, A.; Lewenstam, A.; Maj-Zurawska, M.; Ivaska, A.; Bobacka, J. Comparison of multi-walled carbon nanotubes and poly(3-octylthiophene) as ion-to-electron transducers in all-solid-state potassium ion-selective electrodes. *Electroanalysis* **2011**, *23*, 1352–1358. [CrossRef]
9. Liang, R.; Yin, T.; Qin, W. A simple approach for fabricating solid-contact ion-selective electrode using nanomaterials as transducers. *Anal. Chim. Acta* **2015**, *853*, 291–296. [CrossRef] [PubMed]
10. Lewenstam, A.; Bartoszewicz, B.; Migdalski, J.; Kochan, A. Solid contact reference electrode with a PVC-based composite electroactive element fabricated by 3D printing. *Electrochem. Commun.* **2019**, *109*, 106613. [CrossRef]
11. Bartoszewicz, B.; Dąbrowska, S.; Lewenstam, A.; Migdalski, J. Calibration free solid contact electrodes with two PVC based membranes. *Sensors Actuators B. Chem.* **2018**, *274*, 268–273. [CrossRef]

12. Lingenfelter, P.; Bartoszewicz, B.; Migdalski, J.; Sokalski, T.; Bućko, M.M.; Filipek, R.; Lewenstam, A. Reference electrodes with polymer-based membranes—Comprehensive performance characteristics. *Membranes* **2019**, *9*, 161. [CrossRef] [PubMed]
13. Maksymiuk, K.; Stelmach, E.; Michalska, A. Unintended changes of ion-selective membranes composition—Origin and effect on analytical performance. *Membranes* **2020**, *10*, 266. [CrossRef] [PubMed]
14. Lenar, N.; Paczosa-Bator, B.; Piech, R. Optimization of ruthenium dioxide solid contact in ion-selective electrodes. *Membranes* **2020**, *10*, 182. [CrossRef] [PubMed]
15. Lindner, E.; Guzinski, M.; Pendley, B.; Chaum, E. Plasticized PVC membrane Modified electrodes: Voltammetry of highly hydrophobic compounds. *Membranes* **2020**, *10*, 202. [CrossRef] [PubMed]
16. Lyu, Y.; Gan, S.; Bao, Y.; Zhong, L.; Xu, J.; Wang, W.; Liu, Z.; Ma, Y.; Yang, G.; Niu, L. Solid-contact ion-selective electrodes: Response mechanisms, transducer materials and wearable sensors. *Membranes* **2020**, *10*, 128. [CrossRef] [PubMed]
17. Zajac, M.; Lewenstam, A.; Bednarczyk, P.; Dołowy, K. Measurement of multi ion transport through human bronchial epithelial cell line provides an insight into the mechanism of defective water transport in cystic fibrosis. *Membranes* **2020**, *10*, 43. [CrossRef] [PubMed]
18. Toczylowska-Maminska, R.; Lewenstam, A.; Dolowy, K. Multielectrode bisensor system for time-resolved Monitoring of ion transport across an epithelial cell layer. *Anal. Chem.* **2014**, *86*, 390–394. [CrossRef] [PubMed]
19. Jasielec, J.J.; Filipek, R.; Dołowy, K.; Lewenstam, A. Precipitation of inorganic salts in mitochondrial matrix. *Membranes* **2020**, *10*, 81. [CrossRef] [PubMed]

Publisher's Note: MDPI stays neutral with regard to jurisdictional claims in published maps and institutional affiliations.

© 2020 by the authors. Licensee MDPI, Basel, Switzerland. This article is an open access article distributed under the terms and conditions of the Creative Commons Attribution (CC BY) license (http://creativecommons.org/licenses/by/4.0/).

Article

Reference Electrodes with Polymer-Based Membranes—Comprehensive Performance Characteristics

Peter Lingenfelter [1], Bartosz Bartoszewicz [2], Jan Migdalski [2], Tomasz Sokalski [1], Mirosław M. Bućko [2], Robert Filipek [2] and Andrzej Lewenstam [1,2,*]

[1] Center for Process Analytical Chemistry and Sensor Technology (ProSens), Åbo Akademi University, Piispankatu 8, FI-20500 Turku, Finland; peter.lingenfleter@gmail.com (P.L.); tsokalsk@abo.fi (T.S.)
[2] Faculty of Materials Science and Ceramics, AGH-University of Science and Technology, Al. Mickiewicza 30, PL-30059 Cracow, Poland; bartoszbartosz92@o2.pl (B.B.); migdal@agh.edu.pl (J.M.); bucko@agh.edu.pl (M.M.B.); rof@agh.edu.pl (R.F.)
* Correspondence: andrzej.lewenstam@abo.fi

Received: 10 November 2019; Accepted: 27 November 2019; Published: 29 November 2019

Abstract: Several types of liquid membrane and solid-state reference electrodes based on different plastics were fabricated. In the membranes studied, equitransferent organic (QB) and inorganic salts (KCl) are dispersed in polyvinyl chloride (PVC), polyurethane (PU), urea-formaldehyde resin (UF), polyvinyl acetate (PVA), as well as remelted KCl in order to show the matrix impact on the reference membranes' behavior. The comparison of potentiometic performance was made using specially designed standardized testing protocols. A problem in the reference electrode research and literature has been a lack of standardized testing, which leads to difficulties in comparing different types, qualities, and properties of reference electrodes. Herein, several protocols were developed to test the electrodes' performance with respect to stability over time, pH sensitivity, ionic strength, and various ionic species. All of the prepared reference electrodes performed well in at least some respect and would be suitable for certain applications as described in the text. Most of the reference types, however, demonstrated some weakness that had not been previously highlighted in the literature, due in large part to the lack of exhaustive and/or consistent testing protocols.

Keywords: potentiometry; reference electrode; solid contact; heterogenous membranes; polymer membranes

1. Introduction

Potentiometry, applied routinely in environmental measurements, process analysis, and extensively in clinical chemistry, regained research interest owing to revolutionary developments in electrode design and new materials used. An evident boost to the development activity came with the invention of solid-contact electrodes, which paved the way to all-solid-state electrodes. These electrodes are fully integrated, which allows for different architectures, miniaturization, favorable electrochemical and metrological properties, sterilization, and mass production by unified techniques, such as injection molding or 3D printing [1].

Along with indicator electrodes, their partners in galvanic cells, reference electrodes were developed. As in the case of ion-sensors this was achieved by using ion-to-electron transducing layers, e.g., conducting polymers (CPs) [2].

The reference electrode (RE) is an indispensable and crucial component in potentiometry and open-circuit sensor technology as well as a reference point in amperometric measurements. The failure of the reference electrode means the failure of the entire system. Thus, the quality of the reference electrode is critical in electrochemical measurements, and especially those where multi-parameter analyses are performed. Furthermore, most of the lifetime and size reduction gains from the indicator electrode optimization with CPs are superfluous if the reference electrode cannot be miniaturized in an identical manner.

The classical method of making a reference electrode is by using an electrode of the second kind (Ag/AgCl or Hg/Hg$_2$Cl$_2$) in contact with a solution of a chloride electrolyte, so that the potential determining process is fast and reversible, while the potential itself is stable and reproducible over time, because the composition of the electrolyte is maintained constant. The latter is achieved by placing the electrode in a compartment separate from the sample. This separation, however, allows some electrolytic contact via the salt bridge. To minimize the diffusion potential generated at this junction, the bridge is filled with an equitransferent electrolyte, typically KCl, at a high concentration. Therefore, also in the electrode compartment, the chloride electrolyte is often KCl. A redox electrode (of the 0th kind) in a solution containing a redox couple is sometimes used instead of a RE of the second kind. In principle, this approach is the same as described above.

The quality of the reference electrode is especially important in the direct potentiometric measurement of pH and blood electrolytes where the potential of several indicating electrodes is measured, often in a high-throughput and minimal sample volume automatic analyzer, and where the sample concentrations are measured over a broad range (pH), or in the range where junction potentials are particularly variable (blood electrolytes). In zero-current potentiometry, it is expected that a reference electrode supports reliable measurements. In other words, it is expected that it is sufficiently stable, that it is not fouled by the samples, and that the reference electrode itself does not contaminate the samples. Ideally, it is expected that a reference electrode is easy to manufacture and use, service-free, cheap, and robust.

Although this would seem fairly easy to accomplish, the large body of literature on the subject would argue otherwise [3–20]. The nature of the liquid junction at the salt-bridge plays an especially critical role. When, for instance, the liquid junction becomes clogged, or if the liquid junction is poorly manufactured, errors arise from substantial liquid-junction potentials, which vary with the ionic composition of the solution under test. Optimally, the reference electrode would be of the free-diffusion type or ideally junctionless. The reference junction becomes more and more critical when the ionic strength of the liquid under test is very high or very low, as with, e.g., natural waters and especially power plant water. It has furthermore been shown that calculated residual liquid junctions for these types of samples do not always match what is experimentally exhibited. The contribution of the liquid junction potential arising at the filling solution/sample interface requires often a firm comprehension of potentiometry and possible correction [3,19,20].

Furthermore, miniaturization of liquid-junction-type conventional reference electrodes is difficult, due to the requirement of regular maintenance and a vertical working position. Attempts have been made to develop small REs using the classical approach [21]. The electrolyte (KCl) was placed in a porous material deposited on a flat Ag/AgCl pellet as the substrate. On top of this "immobilized" KCl, the authors placed a film, also porous but with much smaller pores to slow down the release of KCl from the electrode and the contamination of the internal layer with the species from the sample. Obviously, the failure in this approach is lifetime: the smaller the electrode, the faster it is depleted of KCl and contaminated by the sample. Indeed, the authors of [21] reported a 20–90-min lifetime.

Of practical relevance, there are other approaches to the reference electrode, in particular those not utilizing a liquid junction. Since most potentiometric measurements for analytical and thermodynamic purposes are made using cells with liquid junctions, the diffusion potential is observed. A diffusion potential occurs at the boundary between two electrolytes of different composition, where there is a concentration gradient, and thus ion diffusion takes place. Due to different ionic mobilities, some of the ions move faster than others. The different diffusion flows lead to charge separation, and thus an electric field is established. The electric field holds the fast-moving ions and accelerates the slower ions. In the end, a steady state is attained in which equal amounts of the involved ions are transported by a combination of migration and diffusion. Using standard thermodynamic relationships and integrating over the entire boundary region for all species, it can be shown that [22,23]:

$$E_D = \frac{-RT}{F} \int_A^B \sum_i \frac{t_i}{z_i} d \ln a_i \qquad (1)$$

where E_D is the diffusion potential, A and B denote two solutions with different compositions, and t_i, z_i, and a_i are the transference number, charge, and activity of the ith species, respectively, R is the gas constant, T is the temperature, and F is the Faraday constant. It must be noted that this equation is valid regardless of the physical nature of the liquid junction, but, as it involves single ion activities, it cannot be evaluated purely within the framework of thermodynamics. In order to integrate this equation, non-thermodynamic assumptions must be made. The simplest approach mathematically is to assume linear concentration gradients across the liquid junction and constant activity coefficients and ionic mobilities, which results in the well-known Henderson equation [23–25]:

$$E_D = \frac{-RT \sum_i u_i (c_B - c_A)}{F \sum_i u_i z_i (c_B - c_A)} \ln \frac{\sum_i u_i z_i c_B}{\sum_i u_i z_i c_A} \qquad (2)$$

where, A and B denote two solutions with different composition, u_i is the mobility, c_i is the concentration, z_i is the charge, R is the gas constant, T is the temperature, and F is the Faraday constant. Various alternative equations have been derived over the years, but in general, although they may be regarded as more rigorous, no significant differences in calculated diffusion potentials can be expected in most cases.

In direct potentiometry, even small changes in the junction potential can lead to erroneous results, especially when the analytical potential range is narrow. One example is human serum, where the usual concentration ranges for ionized calcium, potassium, and sodium are normally 1.0–1.5, 3.0–6.0, and 120–160 mmol/L, which respectively corresponds to potential intervals of 5.2, 17.8, and 6.4 mV [19]. Furthermore, using Henderson's equation, we can only estimate the diffusion potential. In some cases, this estimation is good enough but in others (human serum) it is questionable. Thus, a few criteria can be formulated with which to evaluate a reference electrode:

(1) The absolute value of the diffusion potential at the liquid junction should be as small as possible.
(2) Changes to the diffusion potential resulting from changes in the sample composition should be as small as possible.
(3) Changes to the diffusion potential resulting from changes to the reference electrode composition should be as small as possible.

These aims are usually achieved by using a high concentration of electrolyte in the bridge of the reference electrode such as 3 M KCl. Such bridges, called hypertonic in biological research, have an intrinsic disadvantage, which is the possibility of denaturation of proteins and/or crystallization of the electrolyte itself. Both may cause instability in the reference electrode potential. The problem may be avoided by using dilute solution in the bridge (isotonic bridge). However, this leads to higher diffusion potentials since isotonic bridges are more sensitive to changes in the sample ionic strength. All of the requirements described make it challenging to build an ideal reference electrode. For the same reason, the number of research papers on reference electrodes is disproportionately small compared to the number of papers on indicator electrodes.

In the past 25 years reference electrode research has begun to regain serious attention [26–70].

Beyond the conventional designs prevalent worldwide, several other compelling approaches have been explored:

i. Equitransferent Salts Dispersed in a Polymer or Other Solid

The challenge of designing functional solid contact reference electrodes was undertaken by Russel who offered a solid-state membrane for reference electrodes made of a polyvinyl resin doped with a very large amount of KCl (1:1 w/w KCl/resin) fabricated under the commercial name, REFEX. The electrochemical characterization of this material was delivered in 1994 [26–28]. Surprisingly, despite the heavy salt loading and large surface area in contact with the liquid sample, the reported leakage of KCl into the sample solution is less than what occurs with conventional ceramic frit junctions. The junction potential is quick to stabilize and relatively constant with time even in media with a very low ionic strength. There are also a number of other papers presenting similar constructions using different polymers or resins, e.g., pressed Al_2O_3-PTFE, urea-formaldehyde, poly(methyl methacrylate)—propylene carbonate, and/or polyester resin [29–33]. An all-solid reference electrode consisting of a sintered Ag/AgCl mixture embedded in solid remelted KCl was as well proposed [34]. Although these concepts are rather different on the surface, the unifying factor is the controlled release of equitransferent salt from either a polymer matrix, dense glass, or ceramic sinter. Variations of all-solid-state reference electrodes with polycrystalline powders of tungsten-substituted alkali molybdenum bronzes mixed with polyester resin were offered [33–35]. These reference electrodes showed no response to changing pH, Na^+ concentration or redox potential. Unfortunately, all of these electrodes showed a relatively high electrical resistance (about 1–500 MΩ), and it was reportedly not possible to get reproducible results.

Recently, the pioneering idea of Russel was extended in the research of Lewenstam's group [70]. Granholm et al. showed that the REs can be produced by dispersing KCl in polypropylene during injection-molding [71], while Mousavi el al. [72] demonstrated that the polymer of a solid-state reference electrode can serve as an embodiment for ion sensors.

Furthermore, it was demonstrated that the PVC heterogenous membranes with silver bromide-KBr salts work superbly as the all-solid-state reference electrodes [73] and that both KBr- and KCl-containing reference membranes are excellent internal solid contact for ion-selective electrodes [74].

These reports provide the signal of a breakthrough in the reference electrode technology which may be called "heterogenous membrane revolution" [73,74]. Material-wise, two aspects of novelty are striking: (1) the application of inorganic salts, and (2) new inert binders which constitute a composite membrane and mechanically processable membranes suitable for electrochemical measurements.

ii. Two Ion-Selective Electrode (ISE) Membranes Connected in Parallel

This concept, while illustrative, is only useful in extremely rare cases and has been thoroughly discussed elsewhere [38,39].

iii. Compensated Cationic and Anionic Response in a Polymer Membrane or Conducting Polymer Film owing to Close-to-Equal Permeability

Ionic liquids [44–49], quaternary ammonium borates [42,50–54,60,61], or other materials [51,52,55,56] are dispersed in, e.g., polyurethane [21,41,55–58], poly(vinyl chloride) (PVC) [42,50,51,60,61], poly(vinyl chloride) carboxylated polymer [21], or polyacrylate [52–54]. Mediating layers such as Nafion are also sometimes applied [40,41,58]. These also variably make use of a conventional inner solution or solid-contact internal materials consisting of conducting polymers, and/or KCl, or NaCl saturated in water, agar, PVC, silicone rubber, mixtures thereof, or other matrices.

An RE with 1-dodecyl-3-methylimidazolium chloride ionic liquid as membrane electrolyte was described in [44]. In fact, this electrode is actually a quasi-reference electrode (QRE) since it works only if the sample contains a high sulfate background. This is not surprising, as it has previously been reported that a similar ionic liquid, 1-butyl-3-methylimidazolium hexafluorophosphate, had been developed into a sulfate ion-selective electrode [62].

Further improvement of many, if not most, of these reference electrodes requires a more detailed knowledge of the functional mechanism. For instance, if the mechanism relies on the distribution potential or on the release of QB from membrane to sample, the use of a solid contact construction instead of the conventional setup with internal filling solution is possible. If, however, the electrodes work due to the release of an inorganic electrolyte (e.g., KCl) extracted from the internal filling matrix (be it liquid or solid), further miniaturization is questionable. It seems unlikely that the latter mechanism would be prevalent, as there have been reports of solid-contact reference electrodes (SCREs) lasting for an extended period of up to two years [21]. However, the stability of one type of SCRE cannot preclude another type of SCRE having another functional mechanism altogether, and therefore each type must in the end be considered separately unless certain general aspects allow for the assumption of a universal functional mechanism.

iv. Polyion-Sensitive ISEs Used as Reference Electrodes

This approach extends interesting concepts introduced with the advent of potentiometric membrane electrodes that are responsive to polyionic analytes [63,64]. The main principle is quite simple: if a membrane electrode can be made responsive to a highly charged analyte, the resulting Nernstian response function will exhibit a very small electrode slope that is inversely proportional to the charge of the analyte. If some amount of this analyte is continuously present at the membrane surface, the resulting potential will be nearly independent of its concentration. Since some well-established anticoagulants (such as heparin) are polyions, and membrane electrodes have been specifically designed to measure such anticoagulants in blood, it seems possible to design reference electrode membranes for use in blood samples. The primary disadvantage of this concept is that highly lipophilic ions may ion-exchange with the polyion in the membrane, thereby increasing its response to small ions.

v. Modified Conducting Polymer Reference Electrodes

This classification comprises pH-buffered, multi-layer and overoxidized junctionless reference electrodes with only electrochemically deposited conducting polymers on the conducting substrate [50,65–68]. A more versatile solution was proposed in [69]. The electrode described contained a layer of a conducting polymer (PEDOT or PMPy) doped with a high concentration of pH buffer. The electrode, in fact, was pH-sensitive, but the sample pH in the vicinity of the electrode was buffered by the electrode itself, thus ensuring a constant potential regardless of the composition of the sample.

Our goal in this work was not to take one variety and improve upon it with the exclusion of all others. Questions of mechanism, miniaturization, etc. are avoided. In this report, we attempt to reproduce and compare electrodes at least similar to those described in [26–28,30,35,55–58,60,61], or as we designated them, REFEX (PVA), urea-formaldehyde (UF), remelted KCl (RKCl), and PVC-(QB(PVC)), and polyurethane-based (QB(PU)) conventional membrane electrodes, where the membranes were loaded with the lipophilic salt, tetrabutyl ammonium tetrabutyl borate (QB).

2. Materials and Methods

For the primary reference electrode, against which all other references were tested, we used a Thermo Orion Ross Ultra double-junction reference electrode purchased from Thermo Fisher Scientific, Waltham, MA, USA. The behaviors of the electrodes were compared with two commercial electrodes: an Orion Ross Sureflow double-junction reference electrode purchased from Thermo Fisher Scientific, Waltham, Massachusetts, USA and a REFEX reference electrode obtained from Refex Sensors Ltd, Westport, C. Mayo, Ireland.

2.1. Chemicals and Materials

Aqueous standard and test solutions were prepared from analytical grade reagents and Elga deionized water (18.2 MΩ·cm). Selectophore®grade high molecular weight poly(vinyl chloride) (PVC), and tetrahydrofuran (THF) were purchased from Sigma Aldrich (Steinheim, Germany). Tetrabutylammonium tetrabutylborate (97%, QB), 2,2-dimethoxy-2-phenylacetophenone (DMPP), vinyl acetate (≥99.0%, VA), poly(vinyl acetate) (M_w ~100,000), silver wire (≥99.9%), potassium chloride (≥99.0%), formaldehyde solution (purum, 37% in water stabilized with 10% methanol), urea (≥99.5%), sodium acetate (≥99%), and DIN19266 pH 4.008, 6.865, 7.413, and 9.180 buffer standard solutions were purchased from Sigma Aldrich (Steinheim, Germany). Tecothane®polyurethane (PU) was obtained from Lubrizol. Air release additive BYK®-A 515 was obtained from Algol Chemicals Oy (Espoo, Finland).

2.2. Preparations

i. Preparation of Basic Cocktails for Membranes

For the purpose of this report, the term "basic cocktail" refers to solutions of PVC or PU, plasticizer, and organic electrolyte (if applicable) in THF. These cocktails are fully transparent homogeneous solutions.

Basic cocktails were prepared as follows: appropriate amounts of organic electrolyte (QB) [60,61] were placed into 4 ml sample vials. The required exact weights of other components were calculated in accordance with the actual weight of the organic electrolyte and then added. The vials were shaken to mix the ingredients and then THF was added. The vials were allowed to spin on a Stuart SRT6 roller mixer overnight to ensure complete dissolution of the PVC or PU. An ultrasound was used when the PVC proved difficult to dissolve. A 15% dry mass was used with the PVC basic cocktails and an 8% dry mas was used with the PU basic cocktails.

ii. Preparation of Ag/AgCl Electrodes

All steps were conducted in a dust-free fume hood.

Washing: Ag (99.9%) pins were first soaked in acetone for 10 min. The pins were placed on lint-free paper and the acetone was allowed to evaporate. The pins were then soaked in HCl (37%) for 20 min to get characteristic metallic color of silver. The HCl was neutralized with NaOH and the pins removed. The pins were rinsed four times with deionized water and again allowed to dry on lint-free paper in a dust-free fume hood. The Ag pins were then inserted and glued into hard PVC caps.

Chloridization: Up to 150 pins were chloridized at one time, using an I-tech IT6322 power supply (IT6322 30V/3A*2CH + 5V/3A*1CH) equipped with three channels. Jigs by which 50 pins could be connected to one of the anode channels of the power source were set on top of the base vessels containing 1 M HCl. Platinum counter electrodes placed in all four corners of each chloridizaton vessel were connected to the cathodes of the power supply. The volume of HCl was adjusted so that all of the Ag below the cap of the pin was submerged. A current of 20 ± 3 mA was applied for 1.5 h. The pins were then rinsed, first with tap water and then carefully with deionized water and then dried overnight. All the pins were checked under a microscope to prove uniform coverage of silver by AgCl. The region close to the plastic cap was covered with a thin layer of Loctite 9483 A&B glue to cover any Ag uncovered with AgCl from having any contact with the sample, and to prevent KCl creep through the cap to the electrical connection above.

2.3. Electrode and Body Types

Two main classifications of REs were produced. Membrane-based reference electrodes were made of two types (QB (PVC) and QB (PU)), using the same body type. The completely solid-state reference electrodes (SSREs) consisted of three types: remelted KCl reference electrodes, urea-formaldehyde (UF) resin reference electrodes, and poly(vinyl acetate) (PVA). All these RE types are schematically shown in Figure 1.

i. Conventional Liquid Contact Membrane-Based REs

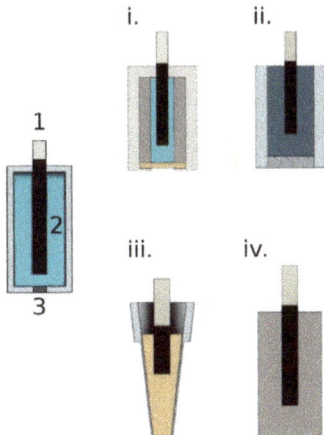

Figure 1. Schematic diagrams of reference electrodes (REs) used. On the left side general scheme of RE, where 1) is the Ag/AgCl electrode, 2) is the internal solution or solid contact, 3) is the membrane or frit. On the right side: (**i.**) RE with QB (PVC) and QB (PU) membranes, (**ii.**) RE with remelted inorganic salts (RKCl), (**iii.**) urea-formaldehyde (UF) resin-based RE, and (**iv.**) RE with PVA membrane. More detailed characterization of (**i.**–**iv.**) type is provided in the text.

Basic membrane cocktails were prepared as described above. 2 mL of the cocktail was poured into a 24.0 mm glass ring mounted onto a glass plate. The ring was covered with a paper and the THF allowed to evaporate for at least 24 h. A cork borer was used to cut 8.0 mm membrane discs from the master membrane. These 8 mm discs were subsequently mounted into Phillips bodies (see: Figure 1i.)

ii. Molten/Remelted KCl Reference Electrode

The electrode body had to withstand high temperatures and mechanical stress during heating. Quartz glass satisfied these requirements. Two quartz frit porosities were tested, with the denser one giving better performance. Potassium chloride was heated to 500 °C to remove the water. Then the body of the electrode was filled with potassium chloride, and the Ag/AgCl wire (2 mm diameter) was suspended in the KCl. The body was placed in the furnace and heated/cooled in three temperature steps of 25 °C → 500 °C → 820 °C → 750 °C (cooling) with the rates 4, 2, and −0.5 °C/min, respectively. The electrode was kept in the furnace until the temperature returned to 25 °C, about 20 h. The silver wire extending out of the KCl was attached to a Metrohm connector model 6.1241.060. (see: Figure 1ii.)

iii. Urea-Formaldehyde Resin + 25–50% w/w KCl Reference Electrodes

The urea-formaldehyde resin was fabricated as described previously [30]. Numerous trials with different mold types resulted in the conclusion that the UF polymer resin did not adhere well to any mold material at our disposal. The poor adhesion allowed the sample to penetrate up the walls of the electrode in an inconsistent fashion. This resulted in some variation in results depending on the surface area of the UF in contact with the sample and tightness of fit to the mold. Limiting the surface area of UF in contact with solution was quite effective in reducing this problem. The best electrodes were those made in 40–200 µL micropipette tips. Wrapping with teflon tape as recommended by the authors in [30] was somewhat effective in reducing penetration of water, but the best results were nonetheless obtained with the micropipette bodies. (see: Figure 1iii.)

iv. Polyvinyl Acetate (PVA) + 50–70% KCl Reference Electrodes

Potassium chloride and sometimes also lithium chloride were ground in a mortar to get a fine powder. The KCl and LiCl were dried for 30 min at 450–500 °C. The powdered salts were mixed with VA and DMPP. The mixture was placed in a mold to form the RE body with the Ag/AgCl wire affixed in the center of the mold. The mold was sealed and placed on a roller mixer (Stuart SRT6) below a 6 W UV lamp (Vilber Lourmat). The mixture was then mixed for 2 min after which it was irradiated using the UV lamp at 365 nm for 40–50 min (with mixing). After irradiation, the form was left for 2 h to cool down. The mold was removed from the hardened mixture by cutting and peeling it off (if the form was made of plastic) or cracking it off (if the form was made of glass). (see: Figure 1iv.)

2.4. Uniform Testing Protocols

In order to objectively compare the behavior and quality of the investigated reference electrodes of different design, uniform testing protocols were devised and adopted. These protocols were developed to test many factors that can influence the reference electrode behavior.

2.4.1. Stability Testing Protocol

The stability criteria arise in two ways. Firstly, the measuring equipment used to evaluate the electrode stability has its own limitations, which we call the 'equipment capability criterion'. The noise from the instrumentation under optimal conditions accounts for approximately 0.005 mV of variation. Secondly, the intended use of the electrode also presents its own set of limitations. These types of limitations give rise to what we call the 'method-based criterion'. If we are measuring in a standard solution after each sample as is often done in clinical analysis, we require only that the drift is small enough that it is negligible in the time frame of two measurements, e.g., 1 min. For instance, if we are measuring Na+ in blood serum, with a normal range of 120–150 mM, with a required coefficient of variation (CV) better than 0.25% [2], the drift could be no more than 0.015 mV/min or 0.87 mV/h as long as it is a stable, regular drift.

On the other hand, as is more typical with, e.g., pH measurements where one would like to measure without a subsequent standard, the reference must remain relatively stable for around one day. In this latter case, we must also specify the intended accuracy of the method over that time span. If we specify an accuracy of 0.1 pH units and assume that the pH electrode itself does not drift, we require a (less stringent) stability of 0.25 mV/h. If we, for example, would rather have a pH test that required calibration only once per week but with 0.2 pH unit accuracy, we would require a (more stringent) stability of approx. 0.05 mV/h.

Potential stability can be described as a property of the electrode to maintain the same potential in certain conditions. It is usually expressed as a standard deviation over a certain period of time. There are two main sources of instability:

Noise, which is a random effect, can be diminished by increasing the number of measurements. It is described as a standard deviation or as a span of the potential. If the noise is expressed as the span, then there is a relation between span and standard deviation as follows: span $\approx 2 \times SD$. There can be many sources of noise, such as electronic equipment, power supply network, the electrode itself, etc.

Drift, which is a systematic effect, might be corrected (bias) during measurements (considering drift via mathematical formulae or conducting calibrations more frequently). It is usually described as a change (shift) of the potential over time.

In real situations, noise and drift occur simultaneously and can be estimated. However, the drift in most cases is beyond interest due to frequent recalibrations, repeated short measurement times or software corrections, whereas the noise is taken into account to estimate uncertainty.

Measurements to investigate the long-term stability of the reference electrodes were carried out in 10^{-4} M KCl solution against the commercial ORION 800500U ROSS Ultra D/J RE. The measurements were carried out in a Faraday cage. The data was recorded and collected using a Lawson Labs 16-channel potentiometer. Measurements were taken every 5 seconds and were obtained using the EMF Suite 2.0 software. The standard deviations for measuring-times over 1, 10, and 20 h were calculated. To better compare the results while still being able to visibly see deviations, the potentials were normalized for 1, 10, and 20 h by taking the first point as "0". The drift was calculated as the linear slope approximation by taking results from the final 30 h. The more stringent tolerance range is represented as two red horizontal dotted lines in the figures. The total potential scale is adjusted to the less stringent criterion, except in Figure 2c, where the range is represented by two black horizontal dotted lines.

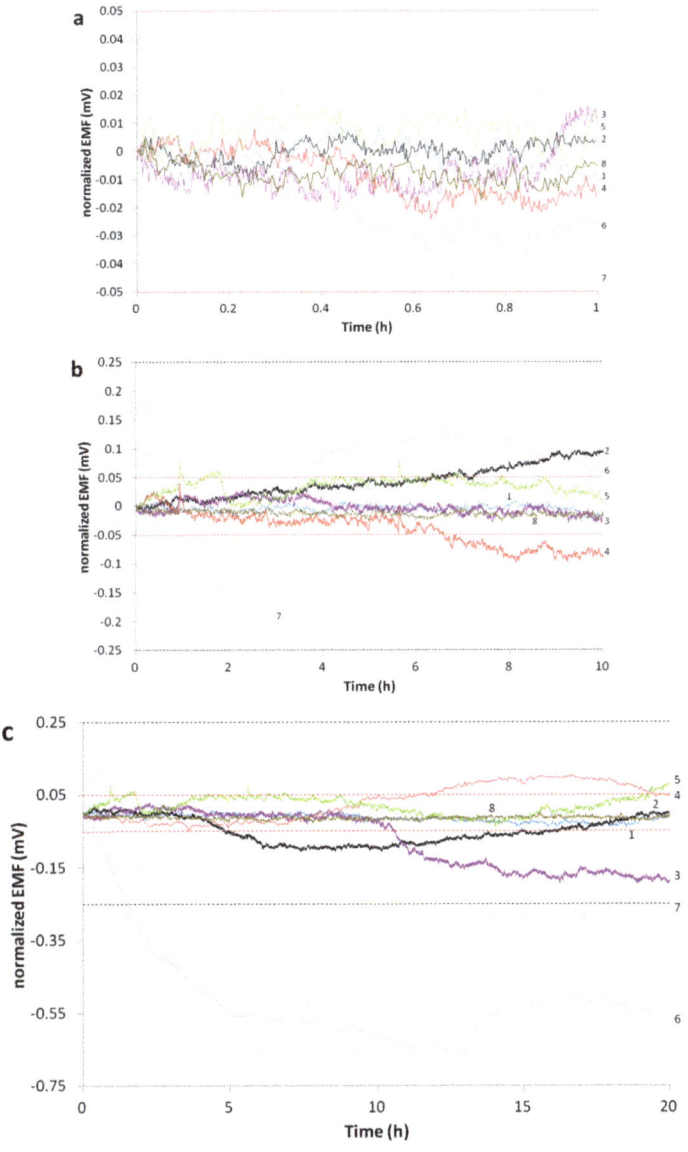

Figure 2. Stability over (**a**) 1 h, (**b**) 10 h, and (**c**) 20 h in 10^{-4} M KCl at 23 °C for 1. QB (PVC), 2. QB (PU), 3. RKCl, 4. PVA, 5. UF, 6. Sureflow, 7. REFEX, 8. Orion Ross Ultra.

2.4.2. pH Testing Protocol

The pH protocol was designed to test the reference electrodes' stability in a relatively wide range of pH samples and buffers for which the pH is well defined (see: Table 1). A literature search was conducted to establish what samples have traditionally been used to test reference electrodes. The software PHREEQCI was used to check these literature values, and good agreement was found in all cases. The electrodes were measured for 5 min in each sample and the reported results are the average of the final 2 min unless the result was not stable during that time interval.

Table 1. pH in different solution used.

Dilute Acids	pH (Pitzer eq) [75–78]
1. 50 mM HCl	1.38
2. 10 mM HCl	2.05
3. 1 mM HCl	3.02
4. 0.1 mM HCl	4.01
5. 0.1 mM HCl + 0.1 M KCl	4.03
6. 0.1 mM HCl + 1 M KCl	4.09
Buffers & Dilute Buffers	-
7. Orion PureWater buffer A	6.97
8. Orion PureWater buffer B	4.10
9. 50 mM potassium hydrogen phthalate	4.01
10. 10 mM potassium hydrogen phthalate	4.12
11. 100 mM HOAc/100 mM NaOAc	4.65
12. 10 mM HOAc/10 mM NaOAc	4.71
13. 25 mM KH_2PO_4/25 mM Na_2HPO_4	6.88
14. 2.5 mM KH_2PO_4/2.5 mM Na_2HPO_4	7.06
15. 10 mM disodium tetraborate	9.18 [79,80] ‡
16. 5 mM disodium tetraborate	9.20 [13], 9.21 [5], 9.19 [79]

‡ PHREEQCI's database did not contain the relevant information for boric acid/borate, so values were found in the literature.

2.4.3. pH Titration Procedure

Solutions of 0.05 M NaOH and 0.005 M HCl were prepared. The NaOH was standardized using potassium hydrogen phthalate, and then used to titrate 100 ml of 0.005 M HCl in which the tested reference electrodes were immersed. In Table 2 calculated pH values are given. The potentials were recorded using a 16-Channel Lawson Lab potentiometer and EMF Suite 2.0 software. The potential was measured for 5 min after each titration step.

Table 2. Calculated pH for pH titration.

NaOH mL	0	3.5	7.5	9	9.5	9.8	9.9	9.95	10.1	10.2	10.5	11	12	15
pH (calc)	2.30	2.50	2.93	3.34	3.64	4.04	4.34	4.64	9.66	9.96	10.35	10.65	10.95	11.34

2.4.4. Multi-Solution Testing Protocol (MSP)

This test aimed at studying the effect of the nature and the concentration of the sample electrolyte. The electrode potentials were recorded in the solutions listed below: KCl 3.0 M, deionized water, NaCl 0.01 M, KCl 0.01 M, HCl 0.01 M, deionized water, KCl 3.0 M, NaCl 0.1 M, KCl 0.1 M, NaBr 0.1 M, $NaHCO_3$ 0.1 M, KOH 0.001 M, HCl 0.01 M, and KCl 3.0 M, deionized water. The EMF readings were recorded for 5 min in each sample and the reported results are the average of the final 2 min. The electrodes were rinsed with deionized water between samples.

3. Results and Discussion

As mentioned above, nowhere has anyone attempted to collect the reference electrodes of interest by the several groups working in this area for comparison testing. Since these references have been designed for different purposes, the tests that have been reported for each type have not been uniform, which makes an independent comparison of them difficult. Such a comparison requires some representative tests to demonstrate which of the references are best for potentiometric tests. Here, we attempt to test in a common fashion the interferences of drift, noise, ionic strength, junction potentials due to ionic mobility differentials, ionic species, pH and buffer species.

3.1. Stability

The first issue with any reference electrode is its stability over time. Different applications exert different stresses upon a reference electrode. In our test regime, we chose to use a dilute 10^{-4} M KCl solution as our sample and monitored the reference potential over a period of several days. Such a low concentration was deemed to be enough of a challenge, especially for electrodes from which KCl is presumed to diffuse out into the sample.

Most of the electrode types we studied performed adequately well to be used for clinical measurements in which an online standard is used for E^0 correction, and for pH measurements where calibrations are performed daily. The former places no strong demand on reference electrode stability other than that it not be particularly noisy, and all of the references measured could be used for such an application. The latter criterion (pH) was more challenging, but for pH measurements demanding only 0.1 pH unit accuracy, again all of the reference electrodes studied would be acceptable, assuming no extremely large pH changes between samples. If the pH measurement accuracy should be greater, or the calibration interval longer, the Orion Ross Ultra, QB (PVC), and QB (PU) references demonstrated superior stability. Figure 2a–c shows the potentials' stability of the various reference electrodes graphed together.

The potentials of the QB (PVC) electrodes: QB30 (PVC)-02, QB30 (PVC)-03, and QB30 (PVC)-11 were quite stable. The standard deviations of QB30 (PVC)-11 for measuring times of 1, 10, and 20 h were 4, 6, and 11 µV respectively, and its drift over the last 30 h of testing was only 1 µV/h. Potential over time for electrode QB30 (PVC)-11 is shown in Figure 2a–c. The QB (PU) electrodes performed similarly well. The QB10 (PU) references showed relatively good stability and drift, but were outperformed in every aspect by the QB25 (PU) for which the lowest drift of all test reference was recorded. The QB25 (PU) references were exceeded only by the Orion Ross Ultra reference electrode in terms of drift. The stability of QB25 (PU)-05 is shown in Figure 2a–c.

The RKCl1 electrode was prepared in a body with a high-density frit (low porosity), whereas RKCl2 electrode with a lower density (higher porosity) frit. The two electrodes showed similar behavior and the same parameters (see Table 3). The standard deviation for measuring times of 1, 10, and 20 h are 7, 42, and 66 µV respectively for the RKCl1 electrode, and 7, 12, and 77 µV respectively for the RKCl2 electrode. Both electrodes drifted 7 µV/h. Long-term time development of the potentials for RKCl1 are shown in Figure 2a–c. The standard potential for the RKCl1 electrode was −221.2 mV and for the RKCl2 it was −221.9 mV against the ORION 800500U ROSS Ultra D/J electrode. These results also showed good reproducibility regarding standard potential as well as stability.

The UFREs were also stable over a long period of time. All of the UFREs tested had a 20 h SD below 0.2 mV, and in a few cases the SD was even lower than 0.1 mV. All of the UFREs also displayed very small drift below 20 µV/h. These electrodes furthermore showed good reproducibility regarding standard potential. The long time development of the potential of the electrode UF20f is shown in Figure 2.

Table 3. Stability data for the REs tested.

Electrode	Stability * [µV]			Drift ** [µV/h]
	1 h SD	10 h SD	20 h SD	
QB30(PVC)-03	35	62	124	3
QB30(PVC)-11	4	6	11	1
QB10(PU)-1	5	29	94	23
QB10(PU)-3	8	28	163	19
QB25(PU)-3	3	35	42	0.8
QB25(PU)-5	3	27	36	0.4
RKCl1	7	12	66	7
RKCl2	7	12	77	7
PVA1	10	29	160	21
UF20e	16	34	42	16
UF20f	15	15	24	9
Orion Ross Ultra	3	4	4	0.02 ≈ 0
Orion Ross Sureflow	122	57	140	95
REFEX©	99	224	195	10

* Stability was calculated as the standard deviation for measuring times 1, 10, and 20 h. ** Drift was calculated as the linear slope approximation from the final 30 h of measurement.

Our own PVA references also displayed good stability. Only one example, PVA1, is listed in Table 3 since the stability test was not performed in its entirety with any other pieces. Its standard deviations were not high, and it drifted little over time. The commercial REFEX© RE was, however, the least stable of the commercial electrodes over time, and was characterized by significant noise.

Not surprisingly, the stability of an Orion Ross Ultra against another Orion Ross Ultra was very good. However, strangely an Orion Ross Sureflow RE against the Orion Ross Ultra performed rather poorly, and worse in terms of stability than many of our test references especially in terms of long-term drift. The test was repeated several times with similar results. The experimental references tested here all performed as well or better than the Orion Sureflow double-junction reference and far better than the REFEX®.

3.2. pH Response

The pH response of the electrodes was tested in two ways. First, the references were exposed to a series of samples and buffer solutions, to monitor deviations from normality in the potential measured against the ORUDJRE. The buffers used are listed in the experimental section and below the x-axis in Figure 3, and included most of the primary and secondary buffers. Beyond testing merely response to pH, this test also looked at the reaction of the reference electrode to the chemical make-up of the buffering agent and to the sample or buffer ionic strength. The ionic strengths in general were still far higher than would be found in natural/environmental samples, although the 0.1 mM HCl approximates acid rain.

For most of the experimental electrodes tested, the ionic strength of the buffer had a larger effect than the buffer composition, as evidenced by the relatively large changes in reference response in the measurements in HCl samples. Most telling is the return to nearly normal response upon addition of 0.1 M KCl, which hardly changes the pH. The QB (PU) 25 electrode is particularly impressive in the measurements after the HCl samples, for which the standard deviation of the measured pH over all samples was 0.2 mV. The Orion Sureflow and REFEX REs were very poor, the latter being so bad that the test with it was discontinued after the acid samples.

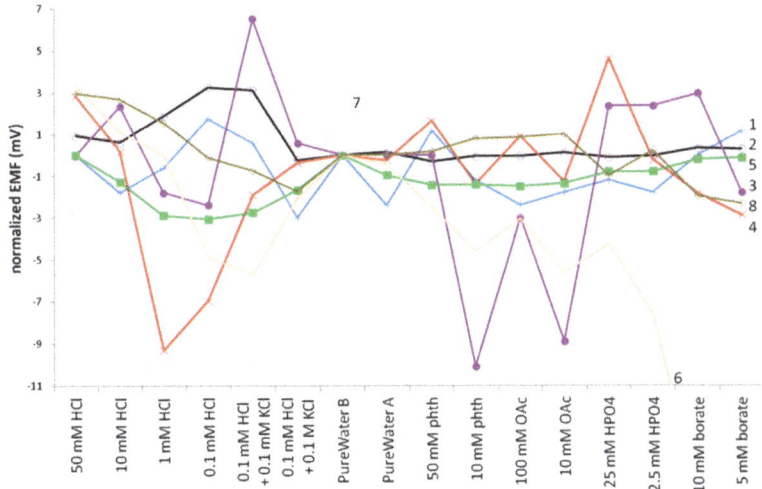

Figure 3. pH titrations with the various reference electrodes tested. 1(+) QB(PVC), 2(o) QB(PU), 3(●) RKCl, 4(×) PVA, 5(■) UF, 6(□) Sureflow, 7(△) REFEX, 8(◇) Orion Ross Ultra.

Second, the electrodes' responses were monitored during a pH titration. As shown in Figure 4, except for once again the REFEX© and perhaps Orion Sureflow references, all of the REs were useful and stable in the pH range 4–10. These results also suggest that, of the experimental REs, the PVA- and UF resin-based REs are the best choices for measurements below pH 4, while a QB (PVC) RE, QB (PU) RE or UFRE is the best above pH 10. It seems that PVAREs are somewhat more sensitive to pH above 10, probably due to hydrolysis of the matrix polymer. In spite of the slight sensitivity, these REs show good parameters in terms of potential stability and reproducibility. The UFREs show even better behavior, being almost insensitive to pH changes. However, in more acidic or alkaline solution some potential deviation does occur. The commercial Orion Ross Ultra shows very good parameters, although it too displays some sensitivity to pH above 11. Assuming both Orion Ross Ultra reference electrodes are identical, this should not have been the case, so presumably one of our Orion Ross Ultra references was partially clogged. Orion recommended procedures to allow freer flow through the liquid junction to remedy this. The REFEX© which is probably based on PVA or another ester resin is very sensitive to pH values, which leads to a rather high SD as shown in Table 4. It should, however, be pointed out that a deviation of 1 mV is equivalent to an error of just under 0.02 pH units, so apart from the REFEX electrode, any of these would be suitable for fairly accurate pH work. No work has been done to characterize the more complicated interactions of these references with temperature and pH.

Table 4. Parameters of the tested REs in the pH titration.

RE	QB (PVC) 30-11	QB (PU) 25-3	QB (PU) 25-5	RKCl	PVA1	PVA6	UF20e	UF20f	Orion Sureflow	Orion Ross Ultra	REFEX
SD [mV]	0.33	0.24	0.44	0.53	0.55	0.26	0.22	0.21	3.8	0.73	12

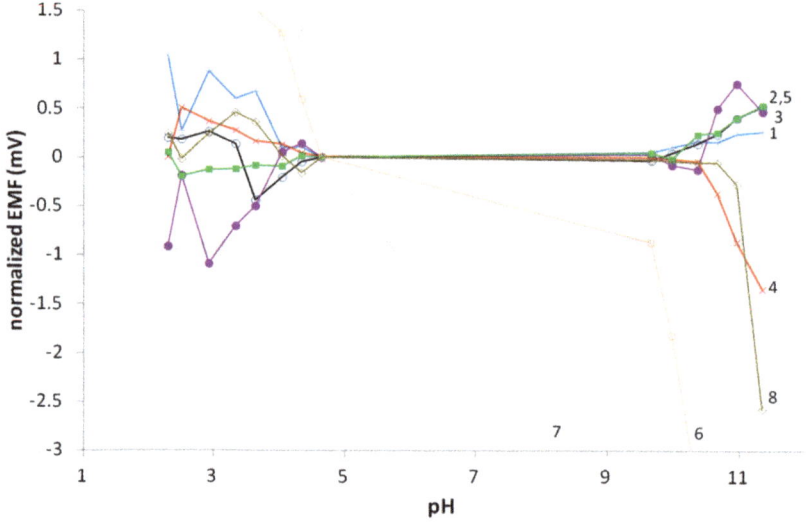

Figure 4. pH titrations with the various reference electrodes tested. 1(+) QB(PVC), 2(o) QB(PU), 3(●) RKCl, 4(×) PVA, 5(■) UF, 6(□) Sureflow, 7(△) REFEX, 8(◇) Orion Ross Ultra.

3.3. Multi-Solution Protocol

The 3.0 M KCl, a number of 0.1 and 0.01 M solutions (described in the paragraph 3.4.4 and represented on the x-axis in Figures 5 and 6), and deionized water were chosen to demonstrate the influence of ionic strength. K^+ and Na^+ demonstrate the influence of the cation, Cl^-, Br^-, and HCO_3^- demonstrate the influence of the anion, while at the same time revealing the effect of disparate ion mobilities. The 0.01 M HCl and 0.001 M KOH show the influence of pH.

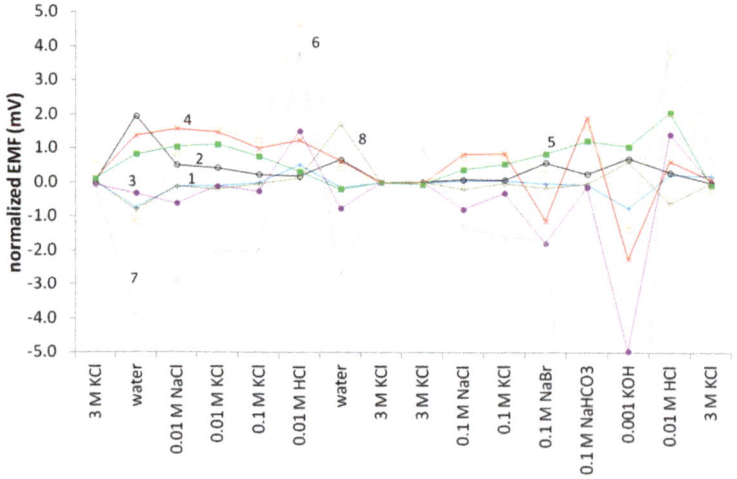

Figure 5. MSP results for the experimental and commercial electrodes tested for this study. 1(+) QB(PVC), 2(o) QB(PU), 3(●) RKCl, 4(×) PVA, 5(■) UF, 6(□) Sureflow, 7(△) REFEX, 8(◇) Orion Ross Ultra.

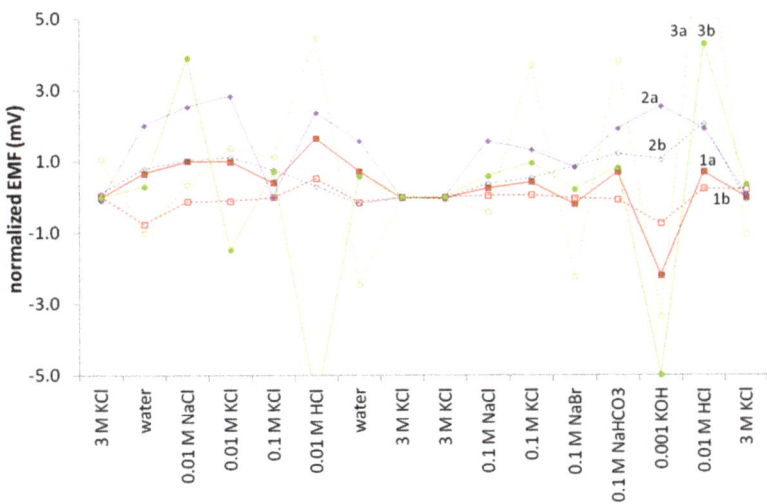

Figure 6. Effect of aging on the 1. QB(PVC), 2. UF, and 3. PVA reference electrodes (**a**) soon after production and (**b**) after three months of semi-regular use.

The order of the electrolytes in the series was designed to ensure the minimization of side effects. In particular, if electrodes drift over time, measurements in a number of 0.01 M solutions and 0.1 M solutions one after another reveal the effect of cation or anion with minimal impact from the drift. Measurements in highly alkaline media are relatively seldom, but on the other hand it is always difficult to wash the electrodes and cell after alkaline solutions. This is why the concentration of hydroxide (0.001 M KOH) was lower than that of other electrolytes, and the measurement in KOH was followed by a measurement in 0.01 HCl.

Figure 5 shows the results for the experimental and commercial REs tested. The electrodes incorporating the lipophilic salt QB performed the best, with little differentiation between the PVC or PU supporting membrane. The other types tested also performed well, with the PVA and UF solid REs slightly better than the remelted KCl REs. The Sureflow commercial reference performed slightly more poorly, and the REFEX commercial RE performed very poorly, especially in the HCl and KOH samples as noticed earlier in the pH testing. The remelted KCl and to a lesser extent the PVA electrodes' performance was influenced by the physical structure and morphology of the electrode bodies. Changing the pH of the sample drastically resulted in generally negative errors in alkaline samples and positive errors in acidic samples because of the glass sinter in the remelted KCl RE bodies and the porosity of the PVA resin. Given enough time (>10–15 min), these errors self-correct but over the short-term errors are recorded. The same effect was reproduced by using a glass sinter body such as that used for the remelted KCl electrodes and filling it with the 20% KCl-UF resin. While its performance generally matched that seen with pipette tip UFREs, the switch from bicarbonate to KOH and then from KOH to HCl resulted in similar errors, as seen with the remelted KCl and PVA REs. Presumably the porosity of the glass sinter was the cause since the same error was not observed with UF micropipette tip REs. The same problem, but greatly exacerbated, is seen with the commercial REFEX electrode.

The age of the electrode also plays an important role, both with traditional types and with any new type, but because of the ratio of the number of electrodes tested to the number of researchers doing the testing, not all of the electrodes were tested uniformly. The QB (PVC) REs were made first, thus allowing for a more thorough investigation of their lifetimes, followed in time order by the solid remelted KCl, PVA, UF electrodes, and finally, the QB (PU) REs. The MSP results for the first three types are shown in Figure 6. The QB (PVC) REs seem to improve somewhat over three months, while the UFREs deteriorated slightly over that same time period. The PVA REs deteriorated more. It is unknown due to the short time frame of the study whether or not these effects are real or due to normal statistical variation.

The problems with UFREs are presumably due to the poor adhesion to every body-type tested except for the glass sinter bodies also used for the remelted KCl REs. It was important to minimize surface area contact between the UF resin and the sample, since a larger surface area in contact with the sample resulted immediately in poor performance. However, even using a micropipette tip as the body, which reduced the surface area to about 0.1 mm^2, over time it was still evident that the sample or conditioning solution had penetrated up the body walls into the electrode. With most pieces, performance deterioration could already be observed after one month of regular use, although some pieces lasted through the duration of the testing (4 months).

3.4. Comments on Manufacturing

Manufacturing of the various types of electrodes was variably challenging. In terms of cost, the remelted KCl electrodes were highest, as the quartz body required for the high-temperature preparation cost EUR 100 alone. The membrane-based reference QB (PVC) and QB (PU) were next in line, since although the membranes are cheap to produce, the bodies into which they are mounted are quite expensive. The solid electrodes were the least expensive, the final price depending on the mass of the electrode, but realistically staying below EUR 1 per electrode.

In terms of complexity of production, the PVA solid references are perhaps most complex due to production requiring distillation, multiple 'ingredients', many weighings, and the final polymerization step using various types of initiator, air release agent, accelerator, etc. Problems with electrode hardness, air bubbles, consistent polymerization of the bulk material, etc. must all be contended with. A switch to using a ready polyester (PE) resin would facilitate the procedure, since one avoids distillation and only two ingredients are needed in addition to the commercial hardener, and to a great extent the problems listed above are also avoided. Unfortunately, initial forays with this approach have not yielded equally good performance. It is likely that the commercial polyvinyl ester resins in use contain proprietary ingredients that are not entirely inert with respect to the tested samples. Remelted KCl is simple once one knows how to deal with the high temperatures. Membrane references are extremely simple for anyone having any experience with ion-selective electrodes.

Tools for reference electrode production are also worth considering. PVA polymerized from the monomer requires distillation apparatus, a mixing system and a UV lamp. UF REs required only a distillation system. Membrane electrodes require almost no hardware, only a cork borer to cut the membranes from the master membrane. Remelted KCl REs require an oven capable of >950 °C.

Polyurethane-based membrane reference electrodes took the longest to make only because the polyurethane takes a few days to dissolve in THF. There is room for improvement in the choice of solvent, but since THF is so ubiquitously used in ISE production, the fact that it can be used with polyurethane is nevertheless a benefit. Otherwise, the PVC-based membrane electrodes require a day for dissolution of the cocktail and another day for membrane casting, and a third day for conditioning. The solid electrodes require some initial hours of preparation and between a few hours to a day for electrode preparation. Conditioning can occur as quickly as in a few minutes, although we normally used a whole day.

The rejection rate of the QB (PVC) references was particularly high, with the percentage rejected between 70–80%. Membranes cut from the same parent membrane and placed into Phillips bodies did not even consistently function identically, indicating either variation in the Phillips bodies or some other uncontrolled variable in the production process. Switching from PVC to PU solved the problem, so the root cause was not pursued. The rejection rate with QB (PU) electrodes was below 20%. With the solid references, rejection rates tended to be lower. The best was perhaps the UF resin-based references in micropipette tips. PVA references did not always work identically. Problems were mostly likely due to centering and depth of the Ag/AgCl wire in the bulk of the solid electrode, air bubbles, and/or hardness after soaking in sample. In a real production process, centering, and depth of the Ag/AgCl pin would be easy to solve. The removal of air bubbles and achieving the desired hardness of the electrode would also not be difficult challenges to overcome with the correct equipment and longer experience. Table 5 collects the production variables for easy comparison.

Table 5. Production variables for the various reference electrode types.

RE Type	Cost Est. *	Complexity /Tooling	Production Time Total (d)/Work (h) †	Rejection Rate	Issues	Sum
QB30(PVC)	200 € §	medium	3/2	80%	rejection	--
QB25(PU)	200 € §	medium	5/2	20%	time to obtain	++
RKCl	120 €	high	1/8	0%	cost	-
PVA	0.37 €	high	1/2	10%	bubbles, hardness, pin placement	+
UF resin	0.43 €	low	1/1	10%	body wall adhesion	+

* Ag/AgCl pin price ignored in all cases; § Phillips body 99.5% of the total price; use of a simplified plastic body drops the price by an order of magnitude. † work time to produce 10 electrodes.

4. Conclusions

The application of a unified set of measurement protocols to multiple reference electrodes allowed us to gain a clearer picture of the shortcomings of the types in question, and to see clearly that these recently reported references are all, in several ways, an improvement over some commercially available but more traditional types of reference electrode. One major shortcoming of this work was the short duration of the testing period which did not give any information concerning the lifetimes of the electrodes. However, the majority of these electrodes are cheap to produce, and even after three months they were mostly superior to two of the commercial electrodes tested.

It is recommended that, in the future work of other groups involved in the development of reference electrodes, tests be conducted at least similar to those described here. Furthermore, it is critical that the mV range on the y-axis of figures be kept as narrow as possible in order that the reader can clearly differentiate whether or not the reference response is really stable. The goal with a potentiometric reference should be stability to within at most a few mV, and when the y-axis is expanded to hundreds of mV, a change of 10 mV is easily hidden by the sheer scale of the figure. Thus, we recommend normalization of the results as much as possible, indicating the potential against, e.g., Ag/AgCl in some other fashion as deemed necessary.

The solid reference types are of great interest because they offer easy fabrication and possibility of dry storage, fast conditioning time, and low cost. A stable readout and low efflux of KCl into the sample depend on morphology. The flexibility of the latter may be offered by the components of the composite and fabrication patterns. In general, to make optimization of these electrodes most efficient, there is a need for a theoretical support of their operation-module.

The clear winner of the empirical studies presented, in terms of performance presented, is the QB(PU) reference electrode. Its stability, pH performance and MSP performance were all the best. The results reported create a rich empirical base which will be used by the authors to elucidate the mechanism of the electrochemical performance of the composite heterogeneous membranes. We are a step further to the point where novel 3D composite membrane structures will revolutionize the world of 1D boundary membranes not only in theory, but in electrode fabrication and application scopes.

Author Contributions: For research articles with several authors, a short paragraph specifying their individual conceptualization, P.L., T.S. and A.L.; methodology, P.L. and A.L.; software, P.L., B.B.; validation, J.M., A.L., M.M.B. and R.F.; formal analysis, A.L.; investigation, P.L., B.B. and J.M.; resources, A.L.; data curation, P.L.; writing—original draft preparation, P.L.; writing—review and editing, A.L., J.M.; visualization, P.L. and B.B.; supervision, A.L.; project administration, A.L.; funding acquisition, A.L.

Funding: National Science Centre (NCN, Poland) financial support via research grant no. 2014/15/B/ST5/02185 is acknowledged.

Acknowledgments: Thanks to Lubrizol for donating polyurethane samples and to Algol Chemicals for BYK®-A 515.

Conflicts of Interest: The authors declare no conflict of interest.

Abbreviations

CP	Conducting Polymer
DMPP	2,2-Dimethoxy-2-Phenylacetophenone
ISE	Ion Selective Electrode
MSP	Multi Solution Protocol
ORUDJRE	Orion Ross Ultra Double Junction Reference Electrode
QB	tetrabutylammonium tetrabutylborate
QB (PVC) RE	Reference Electrode with QB/PVC Membranes
Example: QB30(PVC)-01 means the membrane No 01 with QB content 30% w/w dispersed in PVC	
QB (PU) RE	Reference Electrode with QB/PU Membranes
QRE	Quasi Reference Electrode
PE	Polyester
PEDOT	Poly(3,4-Ethylenedioxythiophene)
PMPy	Poly(1-Methylpyrrole)
PTFE	Poly(Tetrafluoroethylene)
PU	Polyurethane
PVA	Poly(Vinyl Acetate)
PVC	Poly(Vinyl Chloride)
RE	Reference Electrode
RKCl	Remelted KCl
SCs	Solid Contacts
SCRE	Solid Contact Reference Electrode
SD	Standard Deviation
THF	Tetrahydrofuran
UF	Urea Formaldehyde Resin
UFREs	Urea Formaldehyde Resin based Reference Electrode
UV	Ultraviolet Lamp
VA	Vinyl Acetate

References

1. Lewenstam, A. Routines and Challenges in Clinical Application of Electrochemical Ion-Sensors. *Electroanalysis* **2014**, *26*, 1171–1181. [CrossRef]
2. Lewenstam, A. Direct solid contact in reference electrodes. In *Handbook of Reference Electrodes*, 1st ed.; Inzelt, G., Lewenstam, A., Scholz, F., Eds.; Springer-Verlag: Berlin/Heidelberg, Germany, 2013; pp. 279–288.
3. Illingworth, J.A. A common source of error in pH measurement. *Biochem. J.* **1981**, *195*, 259–262. [CrossRef] [PubMed]
4. Brezinski, D.P. Use of half-cell barriers to eliminate junction clogging and thermal hysteresis in silver silver-chloride reference electrodes. *Anal. Chim. Acta* **1982**, *134*, 247–262. [CrossRef]

5. Covington, A.K.; Whalley, P.D.; Davison, W. Recommendations for the determination of pH in low ionic strength fresh waters. *Pure Appl. Chem.* **1985**, *57*, 877–886. [CrossRef]
6. Davison, W.; Woof, C. Performance tests for the measurement of pH with glass electrodes in low ionic strength solutions including natural waters. *Anal. Chem.* **1985**, *57*, 2567–2570. [CrossRef]
7. Davison, W.; Harbinson, T.R. Performance of reference electrodes with free-diffusion junctions: The effect of ionic strength and bore size on junction with simple cylindrical geometry. *Anal. Chim. Acta* **1986**, *187*, 55–65. [CrossRef]
8. Dohner, R.E.; Wegmann, D.; Morf, W.E.; Simon, W. Reference electrode with free-flowing free-diffusion liquid junction. *Anal. Chem.* **1986**, *58*, 2585–2589. [CrossRef]
9. Brennan, C.J.; Peden, M.E. Theory and practice in the electrometric determination of pH in precipitation. *Atmos. Environ.* **1987**, *21*, 901–907. [CrossRef]
10. Davison, W.; Harbison, T.R. Performance of flowing and quiescent free-diffusion junctions in potentiometric measurements at low ionic strengths. *Anal. Chem.* **1987**, *59*, 2450–2456.
11. Davison, W.; Harbinson, T.R. Performance testing of pH electrodes suitable for low ionic strength solutions. *Analyst* **1988**, *113*, 709–713. [CrossRef]
12. Midgley, D. Combination pH electrodes of special design-temperature characteristics and performance in poorly-buffered waters. *Talanta* **1988**, *35*, 447–453. [CrossRef]
13. Davison, W.; Covington, A.K.; Whalley, P.D. Conventional residual liquid junction potentials in dilute solutions. *Anal. Chim. Acta* **1989**, *223*, 441–447. [CrossRef]
14. Ito, S.; Hachiya, H.; Baba, K.; Asano, Y.; Wada, H. Improvement of the silver/silver chloride reference electrode and its application to pH measurement. *Talanta* **1995**, *42*, 1685–1690. [CrossRef]
15. Ito, S.; Kobayashi, F.; Baba, K.; Asano, Y.; Wada, H. Development of long-term stable reference electrode with fluoric resin liquid junction. *Talanta* **1996**, *43*, 135–142. [CrossRef]
16. Peters, G. A reference electrode with free-diffusion liquid junction for electrochemical measurements under changing pressure conditions. *Anal Chem.* **1997**, *69*, 2362–2366. [CrossRef]
17. Ozeki, T.; Tsubosaka, Y.; Nakayama, S.; Ogawa, N.; Kimoto, T. Study of errors determination of hydrogen ion concentrations in rainwater samples using glass electrode method. *Anal. Sci.* **1998**, *14*, 749–756. [CrossRef]
18. Kadis, R.; Leito, I. Evaluation of the residual liquid junction potential contribution to the uncertainty in pH measurement: A case study on low ionic strength natural waters. *Anal. Chim. Acta* **2010**, *664*, 129–135. [CrossRef]
19. Sokalski, T.; Maj-Zurawska, M.; Hulanicki, A.; Lewenstam, A. Optimization of a reference electrode with constrained liquid junction for the measurements of ions. *Electroanalysis* **1999**, *9*, 632–636. [CrossRef]
20. Burnett, R.W.; Covington, A.K.; Fogh-Andersen, N.; Külpmann, W.R.; Lewenstam, A.; Maas, A.H.; Müller-Plathe, O.; VanKessel, A.L.; Zijlstra, W.G. Use of ion-selective electrodes for blood-electrolyte analysis. Recommendations for nomenclature, definitions and conventions. *Clin. Chem. Lab. Med.* **2000**, *38*, 363–370. [CrossRef]
21. Ha, J.; Martin, S.M.; Jeon, Y.; Yoon, I.J.; Brown, R.B.; Nam, H.; Cha, G.S. A polymeric junction membrane for solid-state reference electrodes. *Anal. Chim. Acta* **2005**, *549*, 59–66. [CrossRef]
22. Bard, A.J.; Faulkner, L.R. *Electrochemical Methods*, 1st ed.; John Wiley and Sons, Inc.: New York, NY, USA, 2001.
23. Covington, A.K.; Rebelo, M.J.F. Reference electrodes and liquid junction effects in ion-selective electrode potentiometry. *Ion-Sel. Electrode Rev.* **1983**, *5*, 93–128.
24. Henderson, P.Z. An equation for the calculation of potential difference at any liquid junction boundary. *Z. Phys. Chem. (Leipzig)* **1907**, *59*, 118–127.
25. Hefter, G.T. Calculation of liquid junction potentials for equilibrium studies. *Anal. Chem.* **1982**, *54*, 2518–2524. [CrossRef]
26. Diamond, D.; McEnroe, E.; McCarrick, M.; Lewenstam, A. Evaluation of a new solid-state reference electrode junction material for ion-selective electrodes. *Electroanalysis* **1994**, *6*, 962–971. [CrossRef]
27. Rehm, D.; McEnroe, E.; Diamond, D. An all solid-state reference electrode based on a potassium chloride doped vinyl ester resin. *Anal. Proc.* **1995**, *32*, 319–322. [CrossRef]
28. Desmond, D.; Lane, B.; Alderman, J.; Glennon, J.D.; Diamond, D.; Arrigan, D.W.M. Evaluation of miniaturised solid state reference electrodes on a silicon- based component. *Sens. Actuators B Chem.* **1997**, *44*, 389–396. [CrossRef]

29. Jermann, R.; Tercier, M.-L.; Buffle, J. Pressure insensitive solid state reference electrode for in situ voltammetric measurements in lake water. *Anal. Chim. Acta* **1992**, *269*, 49–58. [CrossRef]
30. Huang, C.L.; Ren, J.J.; Xu, D.F. Study on a new type of all-solid-state reference electrode. *Chin. Chem. Lett.* **1996**, *7*, 1019–1022.
31. Vondrák, J.; Sedlaříková, M.; Velická, J.; Klápště, B.; Novák, V.; Reiter, J. Gel polymer electrolytes based on PMMA: III. PMMA gels containing cadmium. *Electrochim. Acta* **2003**, *48*, 1001–1004. [CrossRef]
32. Reiter, J.; Vondrák, J.; Mička, Z. Solid-state Cd/Cd^{2+} reference electrode based on PMMA gel electrolytes. *Solid State Ion.* **2007**, *177*, 3501–3506. [CrossRef]
33. Gabel, J.; Vonau, W.; Shuk, P.; Guth, U. New reference electrodes based on tungsten-substituted molybdenum bronzes. *Solid State Ion.* **2004**, *169*, 75–80. [CrossRef]
34. Vonau, W.; Olessner, W.; Guth, U.; Henze, J. An all-solid-state reference electrode. *Sens. Actuators B Chem.* **2010**, *144*, 368–373. [CrossRef]
35. Guth, U.; Gerlach, F.; Gerlach, F.; Decker, M.; Olessner, W.; Vonau, W. Solid-state reference electrodes for potentiometric sensors. *J. Solid State Electrochem.* **2009**, *13*, 27–39. [CrossRef]
36. Eine, K.; Kjelstrup, S.; Nagy, K.; Syverud, K. Towards a solid state reference electrode. *Sens. Actuators B Chem.* **1997**, *44*, 381–388. [CrossRef]
37. Nagy, K.; Eine, K.; Syverud, K.; Aune, O. Promising new solid-state reference electrode. *J. Electrochem. Soc.* **1997**, *144*, L1–L2. [CrossRef]
38. Morf, W.; de Rooij, N. Potentiometric response behavior of parallel arrays of cation- and anion-selective membrane electrodes as reference devices. *Electroanalysis* **1997**, *9*, 903–907. [CrossRef]
39. Bakker, E. Hydrophobic membranes as liquid junction-free reference electrodes. *Electroanalysis* **1999**, *11*, 788–792. [CrossRef]
40. Kinlen, P.; Heider, J.E.; Hubbard, D.E. A solid-state pH sensor based on a Nafion-coated iridium oxide indicator electrode and a polymer-based silver chloride reference electrode. *Sens. Actuators B Chem.* **1994**, *22*, 13–25. [CrossRef]
41. Nolan, M.; Tan, S.H.; Kounaves, S.P. Fabrication and characterization of a solid state reference electrode for electroanalysis of natural waters with ultramicroelectrodes. *Anal. Chem.* **1997**, *69*, 1244–1247. [CrossRef]
42. Vincze, A.; Horvai, G. The design of reference electrodes without liquid junction. In *Sensors and Analytical Electrochemical Methods*, 1st ed.; The Electrochemical Society, Inc.: Pennington, NJ, USA, 1997; pp. 550–555.
43. Chudy, M.; Wróblewski, W.; Brzózka, Z. Towards REFET. *Sens. Actuators B Chem.* **1999**, *57*, 47–50. [CrossRef]
44. Maminska, R.; Dybko, A.; Wróblewski, W. All-solid-state miniaturised planar reference electrodes based on ionic liquids. *Sens. Actuators B Chem.* **2006**, *115*, 552–557. [CrossRef]
45. Kakiuchi, T.; Yoshimatsu, T. A new salt bridge based on the hydrophobic room-temperature molten salt. *Bull. Chem. Soc. Jpn.* **2006**, *79*, 1017–1024. [CrossRef]
46. Kakiuchi, T.; Yoshimatsu, T.; Nishi, N. New class of Ag/AgCl electrodes based on hydrophobic Ionic Liquid Saturated with AgCl. *Anal. Chem.* **2007**, *79*, 7187–7191. [CrossRef] [PubMed]
47. Yoshimatsu, T.; Kakiuchi, T. Ionic liquid salt bridge in dilute aqueous solutions. *Anal. Sci.* **2007**, *23*, 1049–1052. [CrossRef] [PubMed]
48. Snook, G.A.; Best, A.S.; Pandolfo, A.G.; Hollenkamp, A.F. Evaluation of a Ag|Ag$^+$ reference electrode for use in room temperature ionic liquids. *Electrochem. Commun.* **2006**, *8*, 1405–1411. [CrossRef]
49. Cicmil, D.; Anastasova, S.; Kavanagh, A.; Diamond, D.; Mattinen, U.; Bobacka, J.; Lewenstam, A.; Radu, A. Ionic liquid-based, liquid-junction-free reference electrode. *Electroanalysis* **2010**, *23*, 1881–1890. [CrossRef]
50. Kisiel, A.; Marcisz, H.; Michalska, A.; Maksymiuk, K. All-solid-state reference electrodes based on conducting polymers. *Analyst* **2005**, *130*, 1655–1662. [CrossRef]
51. Kisiel, A.; Michalska, A.; Maksymiuk, K. Plastic reference electrodes and plastic potentiometric cells with dispersion cast poly(3,4-ethylenedioxythiophene) and poly(vinyl chloride) based membranes. *Bioelectrochemistry* **2007**, *71*, 75–80. [CrossRef]
52. Kisiel, A.; Michalska, A.; Maksymiuk, K.; Hall, E.A.H. All-solid-state reference electrodes with poly(n-butyl acrylate) based membranes. *Electroanalysis* **2008**, *20*, 318–323. [CrossRef]
53. Kisiel, A.; Donten, M.; Mieczkowski, J.; Rius-Ruiz, F.X.; Maksymiuk, K.; Michalska, A. Polyacrylate microspheres composite for all-solid-state reference electrodes. *Analyst* **2010**, *135*, 2420–2425. [CrossRef]

54. Rius-Ruiz, F.X.; Kisiel, A.; Michalska, A.; Maksymiuk, K.; Riu, J.; Rius, F.X. Solid-state reference electrodes based on carbon nanotubes and polyacrylate membranes. *Anal. Bioanal. Chem.* **2011**, *399*, 3613–3622. [CrossRef] [PubMed]
55. Lee, H.J.; Hong, U.S.; Lee, D.K.; Shin, J.H.; Nam, H.; Cha, G.S. Solvent-processible polymer membrane-based liquid junction-free reference electrode. *Anal. Chem.* **1998**, *70*, 3377–3383. [CrossRef]
56. Yoon, H.J.; Shin, J.H.; Lee, S.D.; Nam, H.; Cha, G.S.; Strong, T.D.; Brown, R.B. Solid-state ion sensors with a liquid junction-free polymer membrane-based reference electrode for blood analysis. *Sens. Actuators B Chem.* **2000**, *64*, 8–14. [CrossRef]
57. Nam, H.; Cha, G.S.; Strong, T.D.; Ha, J.; Sim, J.H.; Hower, R.W.; Martin, S.M.; Brown, R.B. Micropotentiometric sensors. *Proc. IEEE* **2003**, *91*, 870–880.
58. Kwon, N.H.; Lee, K.S.; Won, M.S.; Shim, Y.B. An all-solid-state reference electrode based on the layer-by-layer polymer coating. *Analyst* **2007**, *132*, 906–912. [CrossRef] [PubMed]
59. Huang, T.W.; Chou, J.C.; Sun, T.P.; Hsiung, S.K. Fabrication of a screen-printing reference electrode for potentiometric measurement. *Sensor Lett.* **2008**, *6*, 860–863. [CrossRef]
60. Mattinen, U.; Bobacka, J.; Lewenstam, A. Solid-contact reference electrodes based on lipophilic salts. *Electroanalysis* **2009**, *21*, 1955–1960. [CrossRef]
61. Anastasova-Ivanova, S.; Mattinen, U.; Radu, A.; Bobacka, J.; Lewenstam, A.; Migdalski, J.; Danielewski, M.; Diamond, D. Development of miniature all-solid-state potentiometric sensing system. *Sens. Actuators B Chem.* **2010**, *146*, 199–205. [CrossRef]
62. Coll, C.; Labrador, R.H.; Martínez Mañez, R.; Soto, J.; Sancenón, F.; Seguí, M.J.; Sanchez, E. Ionic liquids promote selective responses towards the highly hydrophilic anion sulfate in PVC membrane ion-selective electrodes. *Chem. Commun.* **2005**, 3033–3035. [CrossRef]
63. Mi, Y.M.; Mathison, S.; Bakker, E. Polyion sensors as liquid junction-free reference electrodes. *Electrochem. Solid State Lett.* **1999**, *2*, 198–200. [CrossRef]
64. Meyerhoff, M.E.; Fu, B.; Bakker, E.; Yun, J.H.; Yang, V.C. Peer reviewed: Polyion-sensitive membrane electrodes for biomedical analysis. *Anal. Chem.* **1996**, *68*, 168A–175A. [CrossRef] [PubMed]
65. Mangold, K.M.; Schäfer, S.; Jüttner, K. Reference electrodes based on conducting polymers. *Anal. Chem.* **2000**, *367*, 340–342. [CrossRef] [PubMed]
66. Mangold, K.M.; Schäfer, S.; Jüttner, K. Reference electrodes based on conducting polymer bilayers. *Synth. Met.* **2001**, *119*, 345–346. [CrossRef]
67. Chen, C.C.; Chou, J.C. All-solid-state conductive polymer miniaturized reference electrode. *Jpn. J. Appl. Phys.* **2009**, *48*, 111501. [CrossRef]
68. Ghilane, J.; Hapiot, P.; Bard, A.J. Metal/polypyrrole quasi-reference electrode for voltammetry in nonaqueous and aqueous solutions. *Anal. Chem.* **2006**, *78*, 6868–6872. [CrossRef]
69. Blaz, T.; Migdalski, J.; Lewenstam, A. Junction-less reference electrode for potentiometric measurements obtained by buffering pH in a conducting polymer matrix. *Analyst* **2005**, *130*, 637–643. [CrossRef]
70. Mousavi, Z.; Granholm, K.; Sokalski, T.; Lewenstam, A. An analytical quality solid-state composite reference electrode. *Analyst* **2013**, *138*, 5216–5220. [CrossRef]
71. Granholm, K.; Mousavi, Z.; Sokalski, T.; Lewenstam, A. Analytical quality solid-state composite reference electrode manufactured by injection moulding. *J. Solid State Electrochem.* **2014**, *18*, 607–612. [CrossRef]
72. Mousavi, Z.; Granholm, K.; Sokalski, T.; Lewenstam, A. All-solid-state electrochemical platform for potentiometric measurements. *Sens. Actuators B Chem.* **2015**, *207*, 895–899. [CrossRef]
73. Lewenstam, A.; Blaz, T.; Migdalski, J. All-solid-state reference electrode with heterogeneous membrane. *Anal. Chem.* **2017**, *89*, 1068–1072. [CrossRef]
74. Bartoszewicz, B.; Dąbrowska, S.; Lewenstam, A.; Migdalski, J. Calibration free solid contact electrodes with two PVC based membranes. *Sens. Actuators B Chem.* **2018**, *274*, 268–273. [CrossRef]
75. Pitzer, K.S. Thermodynamics of electrolytes. I. Theoretical basis and general equations. *J. Phys. Chem.* **1973**, *77*, 268–277. [CrossRef]
76. Pitzer, K.S.; Mayorga, G. Thermodynamics of electrolytes. II. Activity and osmotic coefficients for strong electrolytes with one or both ions univalent. *J. Phys. Chem.* **1973**, *77*, 2300–2308. [CrossRef]
77. Pitzer, K.S.; Mayorga, G. Thermodynamics of electrolytes. III. Activity and osmotic coefficients for 2–2 electrolytes. *J. Sol. Chem.* **1974**, *3*, 539–546. [CrossRef]

78. Pitzer, K.S.; Kim, J.J. Thermodynamics of electrolytes. IV. Activity and osmotic coefficients for mixed electrolytes. *J. Am. Chem. Soc.* **1974**, *96*, 5701–5707. [CrossRef]
79. Buck, R.P.; Rondinini, S.; Covington, A.K.; Baucke, F.G.K.; Brett, C.M.A.; Camoes, M.F.; Milton, M.J.T.; Mussini, T.; Naumann, R.; Pratt, K.W.; et al. Measurement of pH Definition, standards, and procedures (IUPAC Recommendations 2002). *Pure Appl. Chem.* **2002**, *74*, 2169–2200. [CrossRef]
80. Camoes, M.F.; Guiomar Lito, M.J.; Ferra, M.I.A.; Covington, A.K. Consistency of pH standard values with the corresponding thermodynamic acid dissociation constants (Technical Report). *Pure Appl. Chem.* **1997**, *69*, 1325–1333. [CrossRef]

© 2019 by the authors. Licensee MDPI, Basel, Switzerland. This article is an open access article distributed under the terms and conditions of the Creative Commons Attribution (CC BY) license (http://creativecommons.org/licenses/by/4.0/).

Review

Unintended Changes of Ion-Selective Membranes Composition—Origin and Effect on Analytical Performance

Krzysztof Maksymiuk, Emilia Stelmach and Agata Michalska *

Faculty of Chemistry, University of Warsaw, Pasteura 1, 02-093 Warsaw, Poland; kmaks@chem.uw.edu.pl (K.M.); ewoznica@chem.uw.edu.pl (E.S.)
* Correspondence: agatam@chem.uw.edu.pl

Received: 29 August 2020; Accepted: 23 September 2020; Published: 28 September 2020

Abstract: Ion-selective membranes, as used in potentiometric sensors, are mixtures of a few important constituents in a carefully balanced proportion. The changes of composition of the ion-selective membrane, both qualitative and quantitative, affect the analytical performance of sensors. Different constructions and materials applied to improve sensors result in specific conditions of membrane formation, in consequence, potentially can result in uncontrolled modification of the membrane composition. Clearly, these effects need to be considered, especially if preparation of miniaturized, potentially disposable internal-solution free sensors is considered. Furthermore, membrane composition changes can occur during the normal operation of sensors—accumulation of species as well as release need to be taken into account, regardless of the construction of sensors used. Issues related to spontaneous changes of membrane composition that can occur during sensor construction, pre-treatment and their operation, seem to be underestimated in the subject literature. The aim of this work is to summarize available data related to potentiometric sensors and highlight the effects that can potentially be important also for other sensors using ion-selective membranes, e.g., optodes or voltammetric sensors.

Keywords: ion-selective membranes; components leakage; incorporation; all-solid-state sensors

1. Introduction

Ion-selective membranes (ISMs) are used in different configurations and in various types of sensors ranging from electrochemical: potentiometric, voltammetric/coulometric, electrolyte gated transistors to optical sensors. An ISM as understood here is a system as typically used in potentiometric sensors: a membrane containing ionophore—a ligand able to preferentially bind analyte ions (called primary ions) in the lipophilic medium [1,2]. Due to the presence of a highly selective ionophore, the ISM allows determination of contents of free ions of interest, in the presence of other chemical forms of analyte, in complex matrices, including blood, serum or environmental samples [3–8]. This makes ISM-based sensors attractive tools for many applications. ISMs are used in various sensor constructions: classical size (macroscopic) and those belonging to nanoscale, intended for disposable use or for long-term operation, containing internal solution or using alternative constructions. Taking into account envisaged applications it is required that sensors are characterized with a high stability of performance, including sensitivity, selectivity as well as stability and reproducibility of potential reading in time [9–11]. It is also advantageous, if the sensor construction allows miniaturization and mass-scale production of devices. The envisaged disposable/in-fields operation sets additional demands such as high reproducibility between different sensors from a production batch, potentially allowing calibration-less operation [12–15].

Classical operation mode of ISM-based sensors is equilibrium mode, ion-exchange occurring between the membrane and the sample is driven by the preference of analyte ions in the ISM phase (typically achieved due to complexation with ionophore) [1,2]; although electrochemical trigger applications (non-equilibrium mode) of ISMs have been also proposed [16]. It is generally accepted that optimal analytical performance of the membrane requires that composition of the phase is well defined by the application of tailored amounts of the defined components during preparation. It is assumed that the intended composition of the membrane is maintained through the sensor's lifetime, with only one exception. In most of cases incorporation of the primary ions (to as prepared membrane) in the pre-treatment step occurs, which is needed to assure stable performance of the sensor.

However, during preparation of potentiometric sensors of different constructions, there are diverse spontaneous processes occurring through the sensor's operation, Figure 1, that may affect analytical responses, the lifetime of the device, or its application safety. Their occurrence can be obscured, and resulting changes can be difficult to trace, leading to variation of performance, that can be attributed to various effects. These processes are generally off the main stream of ISM sensors research, the major focus in the field being on improving performance of the devices. In this work we intend to highlight the processes that can affect ISMs potentiometric sensors operation, and potentially need to be considered while aiming construction improvements, application of new materials etc. It is also shown that considering spontaneous changes and their effect can help to improve sensors, to minimize adverse effects.

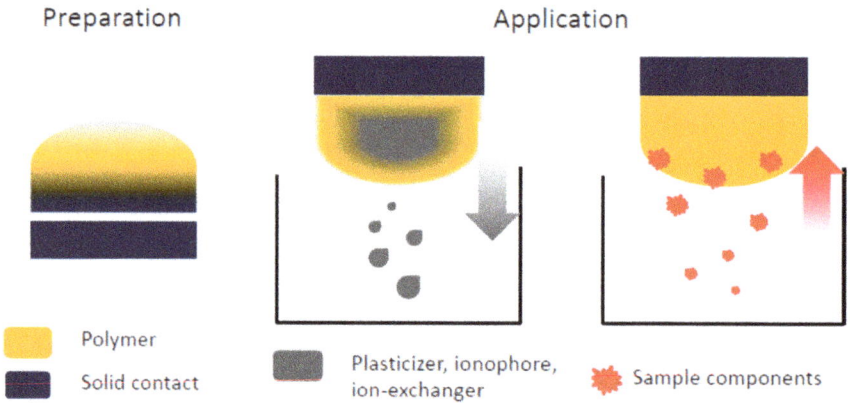

Figure 1. Schematic representation of processes related to ion-selective membrane occurring during sensor preparation as well as application.

2. Ion-Selective Membranes (ISM)

Ion-selective electrodes (ISEs) with polymeric ion-selective membranes have been studied and used for about 50 years [17–22]. The analytical performance of ISMs containing sensors—slope of dependence, linear range of responses, selectivity, are mostly dependent on properties of the ion-selective membrane [1,2,23]. The composition of ISMs is carefully balanced—to contain just the intended compounds of high purity, in the right proportions. Typically, an ISM comprises: ionophore, cation-exchanger, polymer matrix (polymer optionally with plasticizer) [1,2]. Classic composition of an ion-selective membrane is a few *w/w*% (3–5) of each: ionophore and ion-exchanger, and the rest of the membrane mass is the polymer matrix. Typically, the amount of ion-exchanger used is about 60% of the mole amount of ionophore used, thus the presence of excess of free ionophore in the ISM is assured [1,2]. The polymer matrix is mixed with ionophore and ion-exchanger in volatile (auxiliary) solvent, commonly tetrahydrofuran (THF), and applied from dispersion—a membrane cocktail. Spontaneous evaporation of the auxiliary solvent, in the laboratory atmosphere, results in the formation of a polymer layer— the ISM. The membrane can be formed on an inert surface (e.g., glass) and then transferred to the sensor or it can be obtained from a cocktail (solution of membrane components) by drop casting on the top of the transducer of choice to form the sensor. Alternatively, other methods of membrane deposition from a cocktail, e.g., spraying, can be used [24]. If the membrane polymer requires the presence of a plasticizer, which is the most typical case for, e.g., poly(vinyl chloride) (PVC), the compound—liquid used—remains in the formed ISMs (by contrast with THF). For PVC-based systems, plasticizer is relatively abundant in the membrane—its content is typically close to 60% w/w of the film formed.

Some polymeric membranes, e.g., polyacrylates, are formed in a polymerization process from cocktails containing ionophore, ion-exchanger and monomer together with e.g., the cross-linker and polymerization initiator dissolved in monomer solution [25–27]. Under e.g., ultraviolet (UV) irradiation, polymerization of the monomer occurs resulting in film formation. This method of membrane formulation, allowing elimination of liquid plasticizer and auxiliary solvent, seems to be an attractive alternative to avoid problems encountered with classic liquid cocktails of e.g., PVC-based ISMs. Despite the liquid monomers used, the amount of liquid cocktail used is significantly reduced and the application of very good solvent such as THF is eliminated, e.g., [25–27].

It should be stressed that in most cases the ISM is just a physical mixture of components [1,2]. For macroscopic potentiometric-type sensors, the thickness of the membrane is around 150–200 µm, and the membrane is a relatively lipophilic film of a few mm in diameter (typically 5–7 mm), that is in direct contact with the sample at one side, and on the back side it is in contact with the internal solution or solid contact (ion-to-electron transducer) material.

Although the composition of an ISM is precisely controlled during cocktail preparation, formation of the film and its application allows the occurrence of spontaneous processes resulting in change of the membrane composition, ultimately affecting the analytical performance of the sensors. Apart from primary ions exchange with the solution, other components of the ISM and/or solution, can be exchanged across the membrane interfaces: one with the sample or with the internal solution (IS)/solid contact (SC), depending on the construction applied. It can be expected that changes occurring can be driven by the solubility or partition coefficient of the involved species, both in the aqueous phase of the sample and in the organic phase of the membrane during operation and preparation. Leakage of components from the membrane as well as incorporation to the ISM phase can occur, depending on the sample nature, membrane, optionally SC or prevailing conditions. These processes can be related to ISM formation, pre-treatment and operation; being dependent on the construction of the sensor.

Spontaneous processes related to the ISMs operation are rarely considered. The presented reports on spontaneous changes of the composition of ISMs are related to potentiometric sensors mostly; however, it seems rational that these processes will equally involve other systems using similar compositions as ion-selective membranes of voltammetric or optical sensors. Leakage of components from the membrane is increasing in importance if the application of sensors in contact with humans is considered (e.g., wearable sensors, implantable devices) due to related health risks [28]. Moreover, spontaneous changes of ISMs can increase in importance if the volume of the phase is reduced as in the case of membranes of reduced thickness [29] or increased surface to volume ratio, as for e.g., nanospheres optical sensors [30].

In this work we focus on spontaneous effects related to potentiometric sensors ISMs—ion-selective electrodes that are intended to operate in equilibrium conditions. The aim of this work is to highlight the effect of spontaneous processes leading to changes in the composition of ISMs and the potential influence of spontaneous effects on the performance of resulting sensors.

3. Construction of Sensors

The overall performance of ISM-based potentiometric sensors is affected jointly by construction and membrane properties. It is generally assumed that stability of potential readings in time is mainly affected by construction of the sensor [31–34], whereas analytical performance is determined by membrane properties [1,2]. A classic ISM-containing arrangement offers high stability of potential readings in time due to well defined, reversible ion/electron transfer through all sensor interfaces. From the point of view of spontaneous processes related to the ISM this system is clearly less affected by the sensor preparation step. ISMs membranes intended for applications in internal solution ion-selective sensors are usually prepared by applying a membrane components solution—cocktail—to an inert material mould, thus even large quantities of THF used do not lead to accumulation of other substances in the PVC film formed. After solvent evaporation, individual membranes are cut off from the resulting layer and mounted in the sensor housing. Moreover, the internal solution arrangement is typically of well-defined composition and has limited volume (not exceeding 1 mL).

For ISEs, however, miniaturization/mass-scale production is often difficult, thus the solid contact (SC) construction was proposed [31,32,35–37]. The SC arrangement takes advantage of the presence of an ion-to-electron transducer layer between the electron conducting substrate electrode and the ISM. In the SC arrangement, the membrane is in contact with the solid material of various nature/chemical properties, and prepared using different methods [32]. Application of the membrane to make SC-type sensors is principally different from preparation of the ISM for ISE. Due to the variety of materials applied as SC, different conditions prevail during sensor construction. To obtain a membrane, typically an ISM cocktail is drop cast on top of the formed SC—the transducer layer. In this process an auxiliary solvent, e.g., a THF-based solution of the ISM cocktail is in contact with the transducer layer for minutes or hours before the solvent is evaporated. The amount of cocktail applied is dependent on w/w concentration of membrane components and the required thickness of the ISM. Typically, 20–30 ul of membrane cocktail is applied on one substrate electrode of diameter 3 mm [13]. If SC contains material (either main component, additive or impurity) soluble in the cocktail solvent, application of the membrane can result in partition of some of the transducer components to the ISM phase.

Various transducer systems have been proposed, including silver complexes [38], hydrogels [26], redox polymers [39]. For many years conducting polymers (CPs) have remained one of the most popular transducer materials; a wide range of CPs has been tested—from relatively hydrophilic electropolymerized materials like oxidized polypyrrole [27,40,41] or polyaniline [42] to more hydrophobic solvent processable CPs. Among hydrophobic polymers, alkylpolythiohenes, e.g., poly(3-octylthiophene) [43] or composites [44] render hydrophobicity due to the dopant/component applied [45,46], or other systems [47,48]. Although the majority of SC systems proposed use plasticized PVC-based ISMs, other membrane materials. e.g., polyacrylates [27,49] were also successful in this construction.

Electropolymerization of CPs was especially popular in the early years of conducting polymer-based SC systems [31,32]. The benefit of this method of SC formation is well-controlled composition of the formed layer—polymer typically with doping anions [31,32]. This method has been applied for many CP systems that are not solution processable. Typically, highly oxidized and conducting films formed by electropolymerization tend to undergo spontaneous discharge to a more stable oxidized state, and this process is related to exchange of ions with the solution e.g., [50–52]. If discharge of the CP transducer occurs through the ISM, this process can lead to pronounced accumulation of ions at the back side of the membrane, depending on ion-exchange properties of both ISM and transducer layer [53]. Equilibration of the CP before covering with the ISM, although it makes sensor preparation longer [54], helps to avoid excessive ion exchange between the transducer and the membrane, and their consequences such as electrolyte ions accumulation.

It should be stressed that most of transducer layers nowadays are obtained using dispersions e.g., of solvent processable CPs e.g., [55–57]. The transducer material in form of dispersion is e.g., drop cast on the substrate electrode surface, and after evaporation of the solvent it is covered with ISM (in the process similar to that used for electropolymerized SC).

One of the first solution processable materials used as SC were CPs prepared in the presence of a surfactant such as doping anions [58], especially poly(3,4-ethylenedioxythiophene) doped with poly(4-styrenesulfonate) ions (PEDOT–PSS) [55,56], a transducer already successful as a

SC when obtained by electropolymerization [59]. The potential adverse effect of surfactant poly(4-styrenesulfonate) present in the SC was not observed, which can be attributed to its interaction with ions—precipitate formation within the SC, before or after membrane formation, thus preventing its partition to the membrane [56,58].

One of the most popular CPs is polyoctylthiophene (POT) e.g., [57,60–66]. Among advantages of polyalkylthiophenes are solution processability, high lipophilicity and low ions contents in the semiconducting state, which results in a low ion-exchange rate between the SC and ISM. Moreover, POT is soluble in various organic solvents without need of application of surfactants. The possibility of using solution processable materials as solid contacts offers significant advantages in terms of sensor construction, however, it brings significant risk of unwanted, uncontrolled transfer of SC material to the ISM phase, effect that was already mentioned for electropolymerized POT [67,68]. The magnitude of this effect for application of solution-processable POT as transducer is increasing [69]. Application of an ISM cocktail on a formed POT transducer layer results in visible change of the membrane colour and ultimately in CP presence in the membrane in amount close to that of ion-exchanger purposely added, i.e., ca 0.5% w/w [69]. Such high contents of conducting polymer can lead to disturbance in potentiometric responses of the phase, and it needs to be stressed that for sensors based on CP mixed with ionophore, ion-exchanger receptor layers have been proposed previously [70,71].

The other successful group of transducer materials are carbon-based nanostructures: reduced graphene oxide, graphite, macroporous carbon [46,72–77] and especially carbon nanotubes (CNTs) [24,78,79,81]. CNTs, similar to POT, are commercially available and can be prepared as dispersion in solution, allowing application by drop casting [24,78,79,81] or spraying [24]. Sensors using carbon nanostructures as transducer materials are prepared in different variations: using glassy carbon electrodes support, but also on plastic [24,81], or paper [14,82–84]. On the contrary to POT, CNTs are typically applied from dispersions stabilized with surfactant solution, e.g., sodium dodecyl sulfate. In early works of Rius [78,79] about the post-formation of the SC layer, the transducer was washed with water to remove excess of the surfactant. However, the control of effectiveness of this process is not precise. The presence of surfactants affects significantly wettability of CNTs [80,83]. Highly dispersible in water, CNTs containing surfactants are characterized with high hydrophilicity, as estimated using water contact angle [80,83]. Moreover, the presence of unbound surfactants in the transducer layer brings a risk of partition of these compounds to the lipophilic membrane phase, and this effect has been observed previously for samples containing surfactants [85–87]. This in turn can result in impaired performance of the ISM as well as a change of the properties of the transducer and loss of adhesion of membrane phases and ultimately sensor failure [80]. An alternative is dispersing CNTs using other agents, such as POT offering high capacitance of the transducer and high lipophilicity [44], or carboxymethyl cellulose as water-based dispersion. In this case high stability of potentials of prepared sensors was observed [80].

Partition of the transducer material to ion-selective membranes in the course of sensor preparation has been also reported for other systems. Application of SC based on CNTs containing porphyrinoids resulted in spontaneous partition of the latter to the membrane phase [13]. The higher loading of SC with porphyrinoids resulted in higher contents of these in the membrane [13]. It was also clearly confirmed that the presence of porphyrinoids affects performance of the sensor resulting in tailoring fluxes in the ISMs phase and ultimately leading to improved detection limit and selectivity [13].

4. Pretreatment and Operation

Typically, as-prepared ISMs do not contain primary ions and are not equilibrated with an aqueous phase. The only exception is using an ion-exchanger containing a counter ion, the membrane primary ion, e.g., in the case of using potassium salt of cation-exchanger to prepare potassium sensors, sodium or calcium salts to prepare respective sensors [62]. However, even in these cases the prepared ISMs are not equilibrated with water phase, thus fluctuation of recorded potentials can be observed directly after immersing in the sample solution [88]. Pretreatment of the ISM phase results in significant changes in composition of the membrane phase [89], which affects performance of the sensor and ISM analytical parameters [90]. The construction of the sensor applied (IS, or SC) affects this process—SC is typically a more complicated system than IS. For SC, the effect of supporting electrode material as well as properties of the transducer need to be taken into account, also in this step [61,88]. Despite this, some of the processes occurring are similar for IS and SC type sensor.

ISM pretreatment requires analyte ions transfer through the interface between the membrane and solution, formation of a complex with ligand (ionophore) in the membrane, and transport of the formed complex through the membrane. These processes are accompanied by expulsion of ion-exchanger counter ions from the membrane. Depending on composition of the sample and internal solution/solid contact, gradients of ions are formed in the membrane [61,89,91] and ultimately the ion content of the membrane changes. Depending on required sensor performance, pretreatment of ISM can result in different contents of primary ions in the membrane. Sensors intended to show low detection limits [92,93] need to have tailored fluxes of primary and interfering ions in the ISMs phase [94], regardless of construction applied. From a technical/construction point of view, it is typically easier to achieve this for IS type sensors [62,95,96].

For sensors intended to show classical detection limits close to 10^{-6} M, the time needed to spontaneously equilibrate ISMs with solution is dependent on the availability of primary ions in solution and transport of ions in the membrane phase [97,98]. For low sample concentrations (<10^{-4} M) transport of ions in solution becomes the rate-limiting step in the whole equilibration process [99]. However, if the concentration of primary ions in the solution is higher (>10^{-3} M), the rate-limiting step in equilibration of the membrane with primary ions is transport of ions in the membrane phase. Diffusion coefficients of ions in the membrane phase are typically much lower compared to ion-diffusion coefficients in aqueous solution [47,49,89]. Assuming typical thickness of the membrane in the range of 200 µm and diffusion coefficient in the membrane close to 10^{-8} cm^2/s—characteristic for plasticized PVC [89]—the equilibration time needed for high concentration of primary ions in solution is under 12 h. The resulting levels of primary ions are comparable with the ion-exchanger (mole) amount added to the membrane cocktail [97,100]. It should be stressed that the diffusion coefficient obtained for polyacrylate polymers is much lower, close to 10^{-11} cm^2/s, making equilibration with the solution a long-term process [49]. In the case of these materials spontaneous, extremely slow transport of ions within the membrane phase affects significantly the performance of potentiometric sensors [101]. A similar effect, however, in the case of optical sensors results in a significantly increased response range covering even 8 orders of magnitude [102], and offers linear dependence of the signal on logarithm of analyte concentration changes in the sample [71]. One of the possible options to affect (usually slow) ion-diffusion in the ion-selective membrane phase is elimination of the polymer, e.g., using thin liquid layers supported on inert material to host ionophore and ion-exchanger [103,104].

On the other hand, pretreatment of an SC-type sensor can lead also to an ion-exchange process occurring between the membrane and transducer, resulting in a change in composition of this layer. Changes in SC contents have been reported especially for layers originally rich in ions such as dispersion of CPs [61], highly oxidized CPs [53]. This process can lead to advantageous properties of the sensor, e.g., due to binding primary ions within the SC phase coupled with release of interfering ions initially present in SC [58]. It should be stressed that the absolute amount of material used is also important in this respect. Typically, SC contact contains a smaller absolute amount of ions compared to IS, thus it offers limited possibilities of maintaining in the long term the desired ion fluxes through the ISMs. On the other hand, increase of amount of transducer material used to prepare sensors can result in SC being a rich reservoir of ions and ultimately result in unexpected change of performance of the sensor simply due to the change of scale. For example, change of the sensor arrangement from glassy carbon substrate covered with CNTs dispersion to paper type sensors—requiring making conductive paper by application of the CNTs dispersion but in larger quantities—results in alteration of ISE performance. Due to increased contents of interfering ions (originating from dispersion) at the back side of the ISM pronounced ion exchange between the transducer and the membrane is induced [14].

The other issue related to ion-exchange between the ISM and solution is excessive incorporation of primary ions, or coextraction of solution ions. This leads to permselectivity failure [105,106], which is observed mostly in the case of lipophilic ions present in solution as an upper detection limit, or in extreme change of the dependence type from cationic to anionic [107]. The occurrence of these effects has been observed using a potentiometric approach, but also confirmed using spectropotentiometry [108,109] or membrane contents quantification [97,100]. The latter has shown 6 or 8 times higher (mole) amount of primary ions compared to ion-exchanger present in the membrane, for plasticized PVC or polyacrylate more-lipophilic membrane, respectively [100].

It should be stressed that ISMs able to preferentially bind primary ions can lead to accumulation of these ions from the sample, even if the analyte is at the impurities level in the presence of significant excess of other interfering ions [89]. This effect of unintended saturation with primary ions is not readily manifested unless the composition of the membrane is verified with e.g., inductively coupled plasma–mass spectrometry (ICP-MS), as observed in the presence of traces of lead(II) ions ($4 \cdot 10^{-7}$ M) in 0.1 M NaCl solution resulting in accumulation of primary ions in lead-selective membrane [89].

The presence of relatively lipophilic compounds as surfactants, regardless of their charge, in the sample solution, results in deterioration of analytical performance of ISM sensors [85–87]. This effect is attributed to the partition of lipophilic molecules to the membrane phase. Although it was studied for samples and IS-type sensors [85–87], as pointed above this can be observed for SC type sensors, if the SC can be source of surfactants. Presence of lipids in the sample also results in unwanted accumulation of these in the ISM leading to deterioration of performance over time [110].

It has been documented by many reports that pretreatment of the membrane is also connected with incorporation of water to the membrane phase. In polymeric membranes, water transport through the phase is usually characterized with higher diffusion coefficient compared to ion transport. The mechanism of water transport is different from that of ions—water is rather transported through pores of the membrane, whereas ions are moving through the polymeric phase. Water diffusion coefficient of order of 10^{-6} cm^2/s in plasticized PVC membranes has been reported, regardless of used approach ranging from nuclear magnetic resonance (NMR), Fourier transform infrared spectroscopy (FT-IR), holography to electrochemical methods [111–119]. Clearly, water transport through the phase and potential accumulation is dependent on polymer used as matrix, but also transducer used in case of SC.

For SC-type sensors, water transported through ISM can reach the transducer layer. To minimize this risk, it is preferable to use lipophilic materials as the transducer layer. For more hydrophilic solid contacts accumulation of the liquid layer is clearly more probable. Possibility of formation of electrolyte ponds under the ISM phase has been highlighted especially for hydrophilic SC systems as poly(3,4-ethylenedioxythiophene) -poly(styrenesulfonate) [120]. The effect of the presence of an "aqueous layer" within the sensor can manifest itself predominantly in stability of potential readings over time [121,122]. The observed fluctuations of potential can be related to drying/rehydration of the SC layer if the sensor is stored dry/transferred to the sample [61], e.g., can result from just water transport through the phase. However, this process can be coupled with ion-exchange [121], as observed e.g., for ions reaching transducer systems [14], or systems able to bind primary ions within the transducer phase [58].

The presence of water at the back side of the membrane in the case of SC-type sensors can result in chemical change of the electrical lead, used to prepare e.g., simple disposable sensors. This effect has been reported for e.g., screen-printed electrodes used as support for potentiometric sensors—spontaneous chemical change of the printed electrode material, e.g., hydrolysis of inks components resulted in release of the lipophilic species from the SC phase to the membrane, leading to deterioration of sensor performance [72].

Contact of the ISM with the aqueous sample can result in release of membrane components to the solution. These effects have not been widely reported/quantified, nevertheless some of the components can results in toxicological issues, e.g., plasticizer. Recent results have confirmed that plasticizer leakage results in significant contents of these compounds in sample solution [123], reaching e.g., 20 ppm of 2-nitrophenyl octyl ether plasticizer found in sample solution after 12 h contact of this with membrane.

Leaching of ionophore [124] or ion-exchanger [125,126] from the membrane has been observed using an electrochemical approach, looking at change in electrochemical properties of the membrane. These studies clearly confirmed that contents of these components are changing during the sensor's lifetime, this effect can be also related to e.g., absorption of lipophilic species on the membrane surface [124]. Leaching of ion-exchanger from the membrane phase was observed to affect the detection limit of ISM-based potentiometric sensors [127]. It is accepted that spontaneous, unwanted release of ionophore from the membrane limits sensor lifetime [128]. The effect is related to the charge of the ionophore, its lipophilicity and complexation constants [128]. To eliminate unwanted leakage of ionophore, strategies involving immobilization/ covalent attachment were proposed, helping also to prevent other processes occurring between ionophore molecules as e.g., dimerization [98,129–132]. On the other hand, mechanical instability of the sensor as such (e.g., detachment of phases in the case of SC) can also affect performance of potentiometric sensors [80,133,134], and a possible remedy to avoid this effect is covalent binding of sensor phases [135,136].

5. Conclusions

Ion-selective membranes' composition constancy is at continuous risk through sensor construction, pretreatment and operation. Due to the different nature of materials and processes involved, various effects can be observed. More typically, these lead to deterioration of a sensor's performance. Application of new materials to host the ionophore/ion-exchanger to serve as solid contact needs to take into account potential interaction of major and minor components of the used materials with ISM both in preparation and in operation of the sensor.

Funding: Financial support from the National Science Centre (NCN, Poland), project 2018/31/B/ST5/02687, in the years 2019–2022, is gratefully acknowledged.

Conflicts of Interest: The authors declare no conflict of interest.

References

1. Bakker, E.; Bühlmann, P.; Pretsch, E. Carrier-Based Ion-Selective Electrodes and Bulk Optodes. 1. General Characteristics. *Chem. Rev.* **1997**, *97*, 3083–3132. [CrossRef] [PubMed]
2. Bühlmann, P.; Pretsch, E.; Bakker, E. Carrier-Based Ion-Selective Electrodes and Bulk Optodes. 2. Ionophores for Potentiometric and Optical Sensors. *Chem. Rev.* **1998**, *98*, 1593–1688. [CrossRef] [PubMed]
3. Bandodkar, A.J.; Wang, J. Non-invasive wearable electrochemical sensors: A review. *Trends Biotechnol.* **2014**, *32*, 363–371. [CrossRef] [PubMed]
4. Balach, M.M.; Casale, C.H.; Campetelli, A.N. Erythrocyte plasma membrane potential: Past and current methods for its measurement. *Biophys. Rev.* **2019**, *11*, 995–1005. [CrossRef] [PubMed]
5. Justino, C.I.L.; Duarte, A.C.; Rocha-Santos, T.A.P. Critical Overview on the Application of Sensors and Biosensors for Clinical Analysis. *Trends Anal. Chem.* **2016**, *85*, 36–60. [CrossRef] [PubMed]
6. Zhang, Q.; Wang, X.; Decker, V.; Meyerhoff, M.E. Plasticizer-Free Thin-Film Sodium-Selective Optodes Inkjet-Printed on Transparent Plastic for Sweat Analysis. *ACS Appl. Mater. Interfaces* **2020**, *12*, 25616–25624. [CrossRef]
7. De Marco, R.; Clarke, G.; Pejcic, B. Ion-Selective Electrode Potentiometry in Environmental Analysis. *Electroanalysis* **2007**, *19*, 1987–2001. [CrossRef]
8. Zuliani, C.; Diamond, D. Opportunities and challenges of using ion-selective electrodes in environmental monitoring and wearable sensors. *Electrochim. Acta* **2012**, *84*, 29–34. [CrossRef]
9. Zou, X.U.; Cheong, J.H.; Taitt, B.J.; Bühlmann, P. Solid Contact Ion-Selective Electrodes with a Well-Controlled Co(II)/Co(III) Redox Buffer Layer. *Anal. Chem.* **2013**, *85*, 9350–9355. [CrossRef]
10. Zou, X.U.; Zhen, X.V.; Cheong, J.H.; Bühlmann, P. Calibration-Free Ionophore-Based Ion-Selective Electrodes With a Co(II)/Co(III) Redox Couple-Based Solid Contact. *Anal. Chem.* **2014**, *86*, 8687–8692. [CrossRef]

11. Bobacka, J. Potential Stability of All-Solid-State Ion-Selective Electrodes Using Conducting Polymers as Ion-to-Electron Transducers. *Anal. Chem.* **1999**, *71*, 4932–4937. [CrossRef] [PubMed]
12. Zou, X.U.; Chen, L.D.; Lai, C.-Z.; Buhlmann, P. Ionic Liquid Reference Electrodes with a Well-Controlled Co(II)/Co(III) Redox Buffer as Solid Contact. *Electroanalysis* **2015**, *27*, 602–608. [CrossRef]
13. Jaworska, E.; Naitana, M.L.; Stelmach, E.; Pomarico, G.; Wojciechowski, M.; Bulska, E.; Maksymiuk, K.; Paolesse, R.; Michalska, A. Introducing Cobalt(II) Porphyrin/Cobalt(III) Corrole Containing Transducers for Improved Potential Reproducibility and Performance of All-Solid-State Ion-Selective Electrodes. *Anal. Chem.* **2017**, *89*, 7107–7114. [CrossRef] [PubMed]
14. Jaworska, E.; Pomarico, G.; Berna, B.B.; Maksymiuk, K.; Paolesse, R.; Michalska, A. All-solid-state paper based potentiometric potassium sensors containing cobalt(II) porphyrin/cobalt(III) corrole in the transducer layer. *Sens. Actuators B Chem.* **2018**, *277*, 306–311. [CrossRef]
15. Jansod, S.; Wang, L.; Cuartero, M.; Bakker, E. Electrochemical Ion Transfer Mmediated by Alipophilic Os(II)/Os(III) Dinonyl Bipyridyl Probe Incorporated in Thin Film Membranes. *Chem. Commun.* **2017**, *53*, 10757–10760. [CrossRef]
16. Bakker, E. Enhancing ion-selective polymeric membrane electrodes by instrumental control. *TrAC Trends Anal. Chem.* **2014**, *53*, 98–105. [CrossRef]
17. Fiedler, U.; Růžička, J. Selectrode—The Uuniversal Ion-Selective Electrode: Part VII. A Valinomycin-Based Potassium Electrode with Nonporous Polymer Membrane and Solid-State Inner Reference System. *Anal. Chim. Acta* **1973**, *67*, 179–193. [CrossRef]
18. Mascini, M.; Pallozzi, F. Selectivity of neutral carriep-pvc membrane electrodes. *Anal. Chim. Acta* **1974**, *73*, 375–382. [CrossRef]
19. Manning, D.L.; Stokely, J.R.; Magouyrk, D.W. Studies on Several Uranyl Organophosphorus Compounds in a Poly(vinyl chloride) (PVC) Matrix as Ion Sensors for Uranium. *Anal. Chem.* **1974**, *46*, 1116–1119. [CrossRef]
20. Moody, G.J.; Slater, J.M.; Thomas, J.D.R. Poly(vinyl chloride) Matrix Membrane Uranyl Ion-selective Electrodes Based on Organophosphorus Sensors. *Analyst* **1988**, *113*, 669–703.
21. Cammann, K. *Working with Ion.-Selective Electrodes*; Springer Science and Business Media LLC: Berlin, Germany, 1979.
22. Mikhelson, K.N. *Ion-Selective Electrodes*; Springer: Berlin Germany, 2013.
23. Pechenkina, I.A.; Mikhelson, K. Materials for the ionophore-based membranes for ion-selective electrodes: Problems and achievements (review paper). *Russ. J. Electrochem.* **2015**, *51*, 93–102. [CrossRef]
24. Jaworska, E.; Schmidt, M.; Scarpa, G.; Maksymiuk, K.; Michalska, A. Spray-coated all-solid-state potentiometric sensors. *Analyst* **2014**, *139*, 6010–6015. [CrossRef] [PubMed]
25. Heng, L.Y.; Hall, E.A. Methacrylate-acrylate based polymers of low plasticiser content for potassium ion-selective membranes. *Anal. Chim. Acta* **1996**, *324*, 47–56. [CrossRef]
26. Heng, L.Y.; Hall, E.A.H. Producing "Self-Plasticizing" Ion-Selective Membranes. *Anal. Chem.* **2000**, *72*, 42–51. [CrossRef]
27. Michalska, A.; Appaih-Kusi, C.; Heng, L.Y.; Walkiewicz, S.; Hall, E.A.H. An Experimental Study of Membrane Materials and Inner Contacting Layers for Ion-Selective K^+ Electrodes with a Stable Response and Good Dynamic Range. *Anal. Chem.* **2004**, *76*, 2031–2039. [CrossRef]
28. Cánovas, R.; Sanchez, S.P.; Parrilla, M.; Cuartero, M.; Crespo, G.A. Cytotoxicity Study of Ionophore-Based Membranes: Toward On-Body and in Vivo Ion Sensing. *ACS Sens.* **2019**, *4*, 2524–2535. [CrossRef]
29. Cuartero, M.; Acres, R.G.; Bradley, J.; Jarolimova, Z.; Wang, L.; Bakker, E.; Crespo, G.A.; De Marco, R. Electrochemical Mechanism of Ferrocene-Based Redox Molecules in Thin Film Membrane Electrodes. *Electrochim. Acta* **2017**, *238*, 357–367. [CrossRef]
30. Xie, X.; Zhai, J.; Bakker, E. pH Independent Nano-Optode Sensors Based on Exhaustive Ion-Selective Nanospheres. *Anal. Chem.* **2014**, *86*, 2853–2856. [CrossRef]
31. Bobacka, J. Conducting Polymer-Based Solid-State Ion-Selective Electrodes. *Electroanalysis* **2006**, *18*, 7–18. [CrossRef]
32. Michalska, A. All-Solid-State Ion Selective and All-Solid-State Reference Electrodes. *Electroanalysis* **2012**, *24*, 1253–1265. [CrossRef]
33. Hu, J.; Stein, A.; Bühlmann, P. Rational design of all-solid-state ion-selective electrodes and reference electrodes. *TrAC Trends Anal. Chem.* **2016**, *76*, 102–114. [CrossRef]

34. Jarvis, J.M.; Guzinski, M.; Pendley, B.; Lindner, E. Poly(3-octylthiophene) as solid contact for ion-selective electrodes: Contradictions and possibilities. *J. Solid State Electrochem.* **2016**, *20*, 3033–3041. [CrossRef]
35. Shao, Y.; Ying, Y.; Ping, J. Recent advances in solid-contact ion-selective electrodes: Functional materials, transduction mechanisms, and development trends. *Chem. Soc. Rev.* **2020**, *49*, 4405–4465. [CrossRef] [PubMed]
36. Bobacka, J.; Ivaska, A.; Lewenstam, A. Potentiometric Ion Sensors Based on Conducting Polymers. *Electroanalysis* **2003**, *15*, 366–374. [CrossRef]
37. Bobacka, J.; Ivaska, A.; Lewenstam, A. Potentiometric Ion Sensors. *Chem. Rev.* **2008**, *108*, 329–351. [CrossRef]
38. Liu, D.; Meruva, R.K.; Brown, R.B.; Meyerhoff, M.E. Enhancing EMF stability of solid-state ion-selective sensors by incorporating lipophilic silver-ligand complexes within polymeric films. *Anal. Chim. Acta* **1996**, *321*, 173–183. [CrossRef]
39. Hauser, P.C.; Chiang, D.W.; Wright, G.A. A potassium-ion selective electrode with valinomycin based poly(vinyl chloride) membrane and a poly(vinyl ferrocene) solid contact. *Anal. Chim. Acta* **1995**, *302*, 241–248. [CrossRef]
40. Cadogan, A.; Gao, Z.; Lewenstam, A.; Ivaska, A.; Diamond, D. All-solid-state sodium-selective electrode based on a calixarene ionophore in a poly(vinyl chloride) membrane with a polypyrrole solid contact. *Anal. Chem.* **1992**, *64*, 2496–2501. [CrossRef]
41. Michalska, A.; Hulanicki, A.; Lewenstam, A. All Solid-State Hydrogen Ion-Selective Electrode Based on a Conducting Poly(pyrrole) Solid Contact. *Analyst* **1994**, *119*, 2417–2420. [CrossRef]
42. Lindfors, T.; Ivaska, A. Stability of the Inner Polyaniline Solid Contact Layer in All-Solid-State K^+ - Selective Electrodes Based on Plasticized Poly(vinyl chloride). *Anal. Chem.* **2004**, *76*, 4387–4394. [CrossRef]
43. Bobacka, J.; McCarrick, M.; Lewenstam, A.; Ivaska, A. All Solid-State Poly(vinyl chloride) Membrane Ion-Selective Electrodes with Poly(3-octylthiophene) Solid Internal Contact. *Analyst* **1994**, *119*, 1985–1991. [CrossRef]
44. Kałuża, D.; Jaworska, E.; Mazur, M.; Maksymiuk, K.; Michalska, A. Multiwalled Carbon Nanotubes–Poly (3-octylthiophene-2,5-diyl) Nanocomposite Transducer for Ion-Selective Electrodes: Raman Spectroscopy Insight into the Transducer/Membrane Interface. *Anal. Chem.* **2019**, *91*, 9010–9017. [CrossRef] [PubMed]
45. Papp, S.; Bojtár, M.; Gyurcsányi, R.E.; Lindfors, T. Potential Reproducibility of Potassium-Selective Electrodes Having Perfluorinated Alkanoate Side Chain Functionalized Poly(3,4-ethylenedioxythiophene) as a Hydrophobic Solid Contact. *Anal. Chem.* **2019**, *91*, 9111–9118. [CrossRef] [PubMed]
46. Paczosa-Bator, B. Ion-selective electrodes with superhydrophobic polymer/carbon nanocomposites as solid contact. *Carbon* **2015**, *95*, 879–887. [CrossRef]
47. Woźnica, E.; Wojcik, M.; Mieczkowski, J.; Maksymiuk, K.; Michalska, A. Dithizone Modified Gold Nanoparticles Films as Solid Contact for Cu^{2+} Ion-Selective Electrodes. *Electroanalysis* **2012**, *25*, 141–146. [CrossRef]
48. Zeng, X.; Qin, W. A solid-contact potassium-selective electrode with $MoO2$ microspheres as ion-to-electron transducer. *Anal. Chim. Acta* **2017**, *982*, 72–77. [CrossRef]
49. Rzewuska, A.; Wojciechowski, M.; Bulska, E.; Hall, E.A.H.; Maksymiuk, K.; Michalska, A. Composite Polyacrylate–Poly(3,4- ethylenedioxythiophene) Membranes for Improved All-Solid-State Ion-Selective Sensors. *Anal. Chem.* **2008**, *80*, 321–327. [CrossRef]
50. Dumańska, J.; Maksymiuk, K. Studies on Spontaneous Charging/Discharging Processes of Polypyrrole in Aqueous Electrolyte Solutions. *Electroanalysis* **2001**, *13*, 567–573. [CrossRef]
51. Maksymiuk, K. Chemical Reactivity of Polypyrrole and Its Relevance to Polypyrrole Based Electrochemical Sensors. *Electroanalysis* **2006**, *18*, 1537–1551. [CrossRef]
52. Michalska, A.; Ivaska, A.; Lewenstam, A. Modeling Potentiometric Sensitivity of Conducting Polymers. *Anal. Chem.* **1997**, *69*, 4060–4064. [CrossRef]
53. Michalska, A. Improvement of Analytical Characteristic of Calcium Selective Electrode with Conducting Polymer Contact. The Role of Conducting Polymer Spontaneous Charge Transfer Processes and Their Galvanostatic Compensation. *Electroanalysis* **2005**, *17*, 400–407. [CrossRef]
54. Michalska, A.; Hulanicki, A.; Lewenstam, A. All-Solid-State Potentiometric Sensors for Potassium and Sodium Based on Poly(pyrrole) Solid Contact. *Microchem. J.* **1997**, *57*, 59–64. [CrossRef]
55. Michalska, A.; Maksymiuk, K. All-plastic, disposable, low detection limit ion-selective electrodes. *Anal. Chim. Acta* **2004**, *523*, 97–105. [CrossRef]

56. Vázquez, M.; Danielsson, P.; Bobacka, J.; Lewenstam, A.; Ivaska, A. Solution-cast films of poly(3,4-ethylene dioxythiophene) as ion-to-electron transducers in all-solid-state ion-selective electrodes. *Sens. Actuators B Chem.* **2004**, *97*, 182–189. [CrossRef]
57. Michalska, A.; Skompska, M.; Mieczkowski, J.; Zagorska, M.; Maksymiuk, K. Tailoring Solution Cast Poly(3,4-dioctyloxythiophene) Transducers for Potentiometric All-Solid-State Ion-Selective Electrodes. *Electroanalysis* **2006**, *18*, 763–771. [CrossRef]
58. Lindfors, T.; Ivaska, A. All-solid-state calcium-selective electrode prepared of soluble electrically conducting polyaniline and di(2-ethylhexyl)phosphate with tetraoctylammonium chloride as cationic additive. *Anal. Chim. Acta* **2000**, *404*, 111–119. [CrossRef]
59. Vázquez, M.; Bobacka, J.; Ivaska, A.; Lewenstam, A. Small-volume radial flow cell for all-solid-state ion-selective electrodes. *Talanta* **2004**, *62*, 57–63. [CrossRef]
60. McGraw, C.M.; Radu, T.; Radu, A.; Diamond, D. Evaluation of Liquid- and Solid-Contact, Pb^{2+} - Selective Polymer-Membrane Electrodes for Soil Analysis. *Electroanalysis* **2008**, *20*, 340–346. [CrossRef]
61. Michalska, A.; Wojciechowski, M.; Bulska, E.; Maksymiuk, K. Experimental study on stability of different solid contact arrangements of ion-selective electrodes. *Talanta* **2010**, *82*, 151–157. [CrossRef]
62. Michalska, A.; Pyrzyńska, K.; Maksymiuk, K. Method of Achieving Desired Potentiometric Responses of Polyacrylate-Based Ion-Selective Membranes. *Anal. Chem.* **2008**, *80*, 3921–3924. [CrossRef]
63. Rubinova, N.; Chumbimuni-Torres, K.; Bakker, E. Solid-contact potentiometric polymer membrane microelectrodes for the detection of silver ions at the femtomole level. *Sens. Actuators B Chem.* **2006**, *121*, 135–141. [CrossRef]
64. Chumbimuni-Torres, K.Y.; Thammakhet, C.; Galík, M.; Calvo-Marzal, P.; Wu, J.; Bakker, E.; Flechsig, G.-U.; Wang, J. High-Temperature Potentiometry: Modulated Response of Ion-Selective Electrodes During Heat Pulses. *Anal. Chem.* **2009**, *81*, 10290–10294. [CrossRef] [PubMed]
65. Chumbimuni-Torres, K.Y.; Calvo-Marzal, P.; Wang, J.; Bakker, E. Electrochemical Sample Matrix Elimination for Trace-Level Potentiometric Detection with Polymeric Membrane Ion-Selective Electrodes. *Anal. Chem.* **2008**, *80*, 6114–6118. [CrossRef] [PubMed]
66. Chumbimuni-Torres, K.Y.; Rubinova, N.; Radu, A.; Kubota, A.L.T.; Bakker, E. Solid Contact Potentiometric Sensors for Trace Level Measurements. *Anal. Chem.* **2006**, *78*, 1318–1322. [CrossRef] [PubMed]
67. Sutter, J.; Pretsch, E. Response Behavior of Poly(vinyl chloride)- and Polyurethane-Based Ca^{2+} - Selective Membrane Electrodes with Polypyrrole- and Poly(3-octylthiophene)-Mediated Internal Solid Contact. *Electroanalysis* **2006**, *18*, 19–25. [CrossRef]
68. Kim, Y.; Amemiya, S. Stripping Analysis of Nanomolar Perchlorate in Drinking Water with a Voltammetric Ion-Selective Electrode Based on Thin-Layer Liquid Membrane. *Anal. Chem.* **2008**, *80*, 6056–6065. [CrossRef] [PubMed]
69. Jaworska, E.; Mazur, M.; Maksymiuk, K.; Michalska, A. Fate of Poly(3-octylthiophene) Transducer in Solid Contact Ion-Selective Electrodes. *Anal. Chem.* **2018**, *90*, 2625–2630. [CrossRef]
70. Bobacka, J.; Lindfors, T.; McCarrick, M.; Ivaska, A.; Lewenstam, A. Single-piece all-solid-state ion-selective electrode. *Anal. Chem.* **1995**, *67*, 3819–3823. [CrossRef]
71. Kłucińska, K.; Stelmach, E.; Kisiel, A.; Maksymiuk, K.; Michalska, A. Nanoparticles of Fluorescent Conjugated Polymers: Novel Ion-Selective Optodes. *Anal. Chem.* **2016**, *88*, 5644–5648. [CrossRef]
72. Zielińska, R.; Mulik, E.; Michalska, A.; Achmatowicz, S.; Maj-Żurawska, M. All-solid-state planar miniature ion-selective chloride electrode. *Anal. Chim. Acta* **2002**, *451*, 243–249. [CrossRef]
73. Lai, C.-Z.; Fierke, M.A.; Stein, A.A.; Bühlmann, P. Ion-Selective Electrodes with Three-Dimensionally Ordered Macroporous Carbon as the Solid Contact. *Anal. Chem.* **2007**, *79*, 4621–4626. [CrossRef] [PubMed]
74. Jaworska, E.; Lewandowski, W.; Mieczkowski, J.; Maksymiuk, K.; Michalska, A. Critical assessment of graphene as ion-to-electron transducer for all-solid-state potentiometric sensors. *Talanta* **2012**, *97*, 414–419. [CrossRef] [PubMed]
75. Fibbioli, M.; Enger, O.; Diederich, F.; Pretsch, E.; Bandyopadhyay, K.; Liu, S.-G.; Echegoyen, L.; Bühlmann, P. Redox-active self-assembled monolayers as novel solid contacts for ion-selective electrodes. *Chem. Commun.* **2000**, 339–340. [CrossRef]
76. Jaworska, E.; Wójcik, M.; Kisiel, A.; Mieczkowski, J.; Michalska, A. Gold Nanoparticles Solid Sontact for Ion-Selective Electrodes of Highly Stable Potential Readings. *Talanta* **2011**, *85*, 1986–1989. [CrossRef]

77. Górski, Ł.; Matusevich, A.; Pietrzak, M.; Wang, L.; Meyerhoff, M.E.; Malinowska, E. Influence of inner transducer properties on EMF response and stability of solid-contact anion-selective membrane electrodes based on metalloporphyrin ionophores. *J. Solid State Electrochem.* **2008**, *13*, 157–164. [CrossRef] [PubMed]
78. Crespo, G.A.; Macho, S.; Rius, F.X. Ion-Selective Electrodes Using Carbon Nanotubesas Ion-to-Electron Transducers. *Anal. Chem.* **2008**, *80*, 1316–1322. [CrossRef]
79. Crespo, G.A.; Macho, S.; Bobacka, J.; Rius, F.X. Transduction Mechanism of Carbon Nanotubes in Solid-Contact Ion-Selective Electrodes. *Anal. Chem.* **2009**, *81*, 676–681. [CrossRef]
80. Jaworska, E.; Maksymiuk, K.; Michalska, A. Optimizing Carbon Nanotubes Dispersing Agents from the Point of View of Ion-selective Membrane Based Sensors Performance—Introducing Carboxymethylcellulose as Dispersing Agent for Carbon Nanotubes Based Solid Contacts. *Electroanalysis* **2016**, *28*, 947–953. [CrossRef]
81. Michalska, A.; Ocypa, M.; Maksymiuk, K. Highly Selective All-Plastic, Disposable, Cu^{2+} - Selective Electrodes. *Electroanalysis* **2005**, *17*, 327–333. [CrossRef]
82. Novell, M.; Parrilla, M.; Crespo, G.A.; Rius, F.X.; Andrade, F. Paper-Based Ion-Selective Potentiometric Sensors. *Anal. Chem.* **2012**, *84*, 4695–4702. [CrossRef]
83. Mensah, S.T.; Gonzalez, Y.; Calvo-Marzal, P.; Chumbimuni-Torres, K.Y. Nanomolar Detection Limits of Cd^{2+}, Ag^+, and K^+ Using Paper-Strip Ion-Selective Electrodes. *Anal. Chem.* **2014**, *86*, 7269–7273. [CrossRef] [PubMed]
84. Martinez, A.W.; Phillips, S.T.; Butte, M.J.; Whitesides, G.M. Patterned Paper as a Platform for Inexpensive, Low-Volume, Portable Bioassays. *Angew. Chem. Int. Ed.* **2007**, *46*, 1318–1320. [CrossRef] [PubMed]
85. Espadas-Torre, C.; Bakker, E.; Barker, S.; Meyerhoff, M.E. Influence of Nonionic Surfactants on the Potentiometric Response of Hydrogen Ion-Selective Polymeric Membrane Electrodes. *Anal. Chem.* **1996**, *68*, 1623–1631. [CrossRef] [PubMed]
86. Malinowska, E.; Meyerhoff, M.E. Influence of nonionic surfactants on the potentiometric response of ion-selective polymeric membrane electrodes designed for blood electrolyte measurements. *Anal. Chem.* **1998**, *70*, 1477–1488. [CrossRef]
87. Malinowska, E.; Manzoni, A.E.; Meyerhoff, M. Potentiometric response of magnesium-selective membrane electrode in the presence of nonionic surfactants. *Anal. Chim. Acta* **1999**, *382*, 265–275. [CrossRef]
88. Guzinski, M.; Jarvis, J.M.; Pendley, B.; Lindner, E. Equilibration Time of Solid Contact Ion-Selective Electrodes. *Anal. Chem.* **2015**, *87*, 6654–6659. [CrossRef]
89. Michalska, A.; Wojciechowski, M.; Wagner, B.; Bulska, E.; Maksymiuk, K. Laser Ablation Inductively Coupled Plasma Mass Spectrometry Assisted Insight into Ion-Selective Membranes. *Anal. Chem.* **2006**, *78*, 5584–5589. [CrossRef]
90. Bakker, E. Determination of Unbiased Selectivity Coefficients of Neutral Carrier-Based Cation-Selective Electrodes. *Anal. Chem.* **1997**, *69*, 1061–1069. [CrossRef]
91. Schneider, B.; Zwickl, T.; Federer, B.; Pretsch, E.; Lindner, E. Spectropotentiometry: A New Method for in Situ Imaging of Concentration Profiles in Ion-Selective Membranes with Simultaneous Recording of Potential-Time Transients. *Anal. Chem.* **1996**, *68*, 4342–4350. [CrossRef]
92. Sokalski, T.; Ceresa, A.; Fibbioli, M.; Zwickl, T.; Bakker, E.; Pretsch, E. Lowering the Detection Limit of Solvent Polymeric Ion-Selective Membrane Electrodes. 2. Influence of Composition of Sample and Internal Electrolyte Solution. *Anal. Chem.* **1999**, *71*, 1210–1214. [CrossRef]
93. Sokalski, T.; Ceresa, A.; Zwickl, T.; Pretsch, E. Large Improvement of the Lower Detection Limit of Ion-Selective Polymer Membrane Electrodes. *J. Am. Chem. Soc.* **1997**, *119*, 11347–11348. [CrossRef]
94. Gyurcsányi, R.E.; Pergel, É.; Nagy, R.; Kapui, I.; Lan, B.T.T.; Tóth, K.; Bitter, I.; Lindner, E. Direct evidence of ionic fluxes across ion-selective membranes: A scanning electrochemical microscopic and potentiometric study. *Anal. Chem.* **2001**, *73*, 2104–2111. [CrossRef]
95. Michalska, A.; Dumańska, J.; Maksymiuk, K. Lowering the Detection Limit of Ion-Selective Plastic Membrane Electrodes with Conducting Polymer Solid Contact and Conducting Polymer Potentiometric Sensors. *Anal. Chem.* **2003**, *75*, 4964–4974. [CrossRef]
96. Konopka, A.; Sokalski, T.; Michalska, A.; Lewenstam, A.; Maj-Żurawska, M. Factors Affecting the Potentiometric Response of All-Solid-State Solvent Polymeric Membrane Calcium-Selective Electrode for Low-Level Measurements. *Anal. Chem.* **2004**, *76*, 6410–6418. [CrossRef] [PubMed]
97. Woźnica, E.; Wojcik, M.; Wojciechowski, M.; Mieczkowski, J.; Bulska, E.; Maksymiuk, K.; Michalska, A. Improving the Upper Detection Limit of Potentiometric Sensors. *Electroanalysis* **2015**, *27*, 720–726. [CrossRef]

98. Woźnica, E.; Wojcik, M.; Wojciechowski, M.; Mieczkowski, J.; Bulska, E.; Maksymiuk, K.; Michalska, A. Dithizone Modified Gold Nanoparticles Films for Potentiometric Sensing. *Anal. Chem.* **2012**, *84*, 4437–4442. [CrossRef]
99. Michalska, A.; Ocypa, M.; Maksymiuk, K. Effect of interferents present in the internal solution or in the conducting polymer transducer on the responses of ion-selective electrodes. *Anal. Bioanal. Chem.* **2006**, *385*, 203–207. [CrossRef]
100. Michalska, A.; Wojciechowski, M.; Bulska, E.; Maksymiuk, K. Quantifying Primary Silver Ions Contents in Poly(vinyl chloride) and Poly(n-butyl acrylate) Ion-Selective Membranes. *Electroanalysis* **2009**, *21*, 1931–1938. [CrossRef]
101. Woźnica, E.; Mieczkowski, J.; Michalska, A. Electrochemical evidences and consequences of significant differences in ions diffusion rate in polyacrylate-based ion-selective membranes. *Analyst* **2011**, *136*, 4787–4793. [CrossRef]
102. Woźnica, E.; Maksymiuk, K.; Michalska, A. Polyacrylate Microspheres for Tunable Fluorimetric Zinc Ions Sensor. *Anal. Chem.* **2014**, *86*, 411–418. [CrossRef]
103. Baranowska-Korczyc, A.; Jaworska, E.; Strawski, M.; Paterczyk, B.; Maksymiuk, K.; Michalska, A. A Potentiometric Sensor Based on Modified Electrospun PVDF Nanofibers–Towards 2D Ion-Selective Membranes. *Analyst* **2020**, *145*, 5594–5602. [CrossRef] [PubMed]
104. Baranowska-Korczyc, A.; Maksymiuk, K.; Michalska, A. Electrospun nanofiber supported optodes: Scaling down the receptor layer thickness to nanometers – towards 2D optodes. *Analyst* **2019**, *144*, 4667–4676. [CrossRef] [PubMed]
105. Bühlmann, P.; Amemiya, S.; Yajima, S.; Umezawa, Y. Co-Ion Interference for Ion-Selective Electrodes Based on Charged and Neutral Ionophores: A Comparison. *Anal. Chem.* **1998**, *70*, 4291–4303. [CrossRef]
106. Lindner, E.; Zwickl, T.; Bakker, E.; Lan, B.T.T.; Tóth, K.; Pretsch, E. Spectroscopic in Situ Imaging of Acid Coextraction Processes in Solvent Polymeric Ion-Selective Electrode and Optode Membranes. *Anal. Chem.* **1998**, *70*, 1176–1181. [CrossRef]
107. Mathison, S.; Bakker, E. Effect of Transmembrane Electrolyte Diffusion on the Detection Limit of Carrier-Based Potentiometric Ion Sensors. *Anal. Chem.* **1998**, *70*, 303–309. [CrossRef]
108. Gyurcsányi, R.E.; Lindner, E. Spectroscopic Method for the Determination of the Ionic Site Concentration in Solvent Polymeric Membranes and Membrane Plasticizers. *Anal. Chem.* **2002**, *74*, 4060–4068. [CrossRef]
109. Gyurcsányi, R.E.; Lindner, E. Spectroelectrochemical Microscopy: Spatially Resolved Spectroelectrochemistry of Carrier-Based Ion-Selective Membranes. *Anal. Chem.* **2005**, *77*, 2132–2139. [CrossRef]
110. Bühlmann, P.; Hayakawa, M.; Ohshiro, T.; Amemiya, S.; Umezawa, Y. Influence of Natural, Electrically Neutral Lipids on the Potentiometric Responses of Cation-Selective Polymeric Membrane Electrodes. *Anal. Chem.* **2001**, *73*, 3199–3205. [CrossRef]
111. Li, Z.; Li, X.; Petrović, S.; Harrison, D.J. Dual-Sorption Model of Water Uptake in Poly(vinyl chloride)-Based Ion-Selective Membranes: Experimental Water Concentration and Transport Parameters. *Anal. Chem.* **1996**, *68*, 1717–1725. [CrossRef]
112. Chan, A.D.C.; Harrison, D.J. NMR study of the state of water in ion-selective electrode membranes. *Anal. Chem.* **1993**, *65*, 32–36. [CrossRef]
113. Chan, A.D.C.; Li, X.; Harrison, D.J. Evidence for a water-rich surface region in poly(vinyl chloride)-based ion-selective electrode membranes. *Anal. Chem.* **1992**, *64*, 2512–2517. [CrossRef]
114. Sundfors, F.; Lindfors, T.; Höfler, L.; Bereczki, R.; Gyurcsányi, R.E. FTIR-ATR Study of Water Uptake and Diffusion through Ion-Selective Membranes Based on Poly(acrylates) and Silicone Rubber. *Anal. Chem.* **2009**, *81*, 5925–5934. [CrossRef]
115. Appiah-Kusi, C.; Kew, S.J.; Hall, E. Water Transport in Poly(n-butyl acrylate) Ion-Selective Membranes. *Electroanalysis* **2009**, *21*, 1992–2003. [CrossRef]
116. Lindfors, T.; Höfler, L.; Jágerszki, G.; Gyurcsányi, R.E. Hyphenated FT-IR-Attenuated Total Reflection and Electrochemical Impedance Spectroscopy Technique to Study the Water Uptake and Potential Stability of Polymeric Solid-Contact Ion-Selective Electrodes. *Anal. Chem.* **2011**, *83*, 4902–4908. [CrossRef] [PubMed]
117. Lindfors, T.; Sundfors, F.; Höfler, L.; Gyurcsanyi, R.E. FTIR-ATR Study of Water Uptake and Diffusion Through Ion-Selective Membranes Based on Plasticized Poly(vinyl chloride). *Electroanalysis* **2009**, *21*, 1914–1922. [CrossRef]

118. He, N.; Lindfors, T. Determination of Water Uptake of Polymeric Ion-Selective Membranes with the Coulometric Karl Fischer and FT-IR-Attenuated Total Reflection Techniques. *Anal. Chem.* **2012**, *85*, 1006–1012. [CrossRef]
119. Kalinichev, A.V.; Solovyeva, E.V.; Ivanova, A.R.; Khripoun, G.A.; Mikhelson, K. Non-constancy of the bulk resistance of ionophore-based Cd2+-selective electrode: A correlation with the water uptake by the electrode membrane. *Electrochim. Acta* **2020**, *334*, 135541. [CrossRef]
120. Veder, J.-P.; Patel, K.; Clarke, G.; Grygolowicz-Pawlak, E.; Silvester, D.S.; De Marco, R.; Pretsch, E.; Bakker, E.; Silvester, D.S. Synchrotron Radiation/Fourier Transform-Infrared Microspectroscopy Study of Undesirable Water Inclusions in Solid-Contact Polymeric Ion-Selective Electrodes. *Anal. Chem.* **2010**, *82*, 6203–6207. [CrossRef]
121. Cha, G.S.; Liu, D.; Meyeroff, M.E.; Cantor, H.C.; Midgley, A.R.; Goldberg, H.D.; Brown, R.B. Electrochemical Performance, Biocompatibility, and Adhesion of New Polymer Matrices for Solid-State Ion Sensors. *Anal. Chem.* **1991**, *63*, 1666–1672. [CrossRef]
122. Fibbioli, M.; Morf, W.E.; Badertscher, M.; de Rooij, N.; Pretsch, E. Potential Drifts of Solid-Contacted Ion-Selective Electrodes Due to Zero-Current Ion Fluxes Through the Sensor Membrane. *Electroanalysis* **2000**, *12*, 1286–1292. [CrossRef]
123. Kisiel, A.; Kaluza, D.; Paterczyk, B.; Maksymiuk, K.; Michalska, A. Quantifying plasticizer leakage from ion-selective membranes—A nanosponge approach. *Analyst* **2020**, *145*, 2966–2974. [CrossRef] [PubMed]
124. Pendley, B.D.; Lindner, E. A Chronoamperometric Method to Estimate Ionophore Loss from Ion-Selective Electrode Membranes. *Anal. Chem.* **1999**, *71*, 3673–3676. [CrossRef] [PubMed]
125. Pendley, B.D.; Gyurcsányi, R.E.; Buck, R.P.; Lindner, E. A Chronoamperometric Method To Estimate Changes in the Membrane Composition of Ion-Selective Membranes. *Anal. Chem.* **2001**, *73*, 4599–4606. [CrossRef]
126. Paczosa-Bator, B.; Piech, R.; Lewenstam, A. Determination of the leaching of polymeric ion-selective membrane components by stripping voltammetry. *Talanta* **2010**, *81*, 1003–1009. [CrossRef] [PubMed]
127. Telting-Diaz, M.; Bakker, E. Effect of Lipophilic Ion-Exchanger Leaching on the Detection Limit of Carrier-Based Ion-Selective Electrodes. *Anal. Chem.* **2001**, *73*, 5582–5589. [CrossRef] [PubMed]
128. Bühlmann, P.; Umezawa, Y.; Rondinini, S.; Vertova, A.; Pigliucci, A.; Bertesago, L. Headspace Solid-Phase Microextraction. *Anal. Chem.* **2000**, *72*, 1843–1852. [CrossRef] [PubMed]
129. Radu, A.; Peper, S.; Bakker, E.; Diamond, D. Guidelines for Improving the Lower Detection Limit of Ion-Selective Electrodes: A Systematic Approach. *Electroanalysis* **2007**, *19*, 144–154. [CrossRef]
130. Qin, Y.; Bakker, E. Elimination of Dimer Formation in In^{III}Porphyrin-Based Anion-Selective Membranes by Covalent Attachment of the Ionophore. *Anal. Chem.* **2004**, *76*, 4379–4386. [CrossRef]
131. Wang, L.; Meyerhoff, M.E. Polymethacrylate polymers with appended aluminum(III)-tetraphenylporphyrins: Synthesis, characterization and evaluation as macromolecular ionophores for electrochemical and optical fluoride sensors. *Anal. Chim. Acta* **2008**, *611*, 97–102. [CrossRef]
132. Liu, Y.; Xue, Y.; Tang, H.; Wang, M.; Qin, Y. Click-immobilized K+-selective ionophore for potentiometric and optical sensors. *Sens. Actuators B Chem.* **2012**, *171*, 556–562. [CrossRef]
133. Qin, Y.; Peper, S.; Bakker, E. Plasticizer-Free Polymer Membrane Ion-Selective Electrodes Containing a Methacrylic Copolymer Matrix. *Electroanalysis* **2002**, *14*, 1375–1381. [CrossRef]
134. Du, E.; Qiang, Y.; Liu, J. Erythrocyte Membrane Failure by Electromechanical Stress. *Appl. Sci.* **2018**, *8*, 174. [CrossRef] [PubMed]
135. Grygołowicz-Pawlak, E.; Palys, B.; Biesiada, K.; Olszyna, A.R.; Malinowska, E. Covalent binding of sensor phases—A recipe for stable potentials of solid-state ion-selective sensors. *Anal. Chim. Acta* **2008**, *625*, 137–144. [CrossRef] [PubMed]
136. Ocaña, C.; Abramova, N.; Bratov, A.; Lindfors, T.; Bobacka, J. Calcium-selective electrodes based on photo-cured polyurethane-acrylate membranes covalently attached to methacrylate functionalized poly(3,4-ethylenedioxythiophene) as solid-contact. *Talanta* **2018**, *186*, 279–285. [CrossRef]

© 2020 by the authors. Licensee MDPI, Basel, Switzerland. This article is an open access article distributed under the terms and conditions of the Creative Commons Attribution (CC BY) license (http://creativecommons.org/licenses/by/4.0/).

Article

Optimization of Ruthenium Dioxide Solid Contact in Ion-Selective Electrodes

Nikola Lenar, Beata Paczosa-Bator * and Robert Piech

Faculty of Materials Science and Ceramics, Mickiewicza 30, AGH University of Science and Technology, PL-30059 Krakow, Poland; nlenar@agh.edu.pl (N.L.); rpiech@agh.edu.pl (R.P.)
* Correspondence: paczosa@agh.edu.pl

Received: 8 July 2020; Accepted: 7 August 2020; Published: 9 August 2020

Abstract: Ruthenium dioxide occurs in two morphologically varied structures: anhydrous and hydrous form; both of them were studied in the scope of this work and applied as mediation layers in ion-selective electrodes. The differences between the electrochemical properties of those two materials underlie their diverse structure and hydration properties, which was demonstrated in the paper. One of the main differences is the occurrence of structural water in $RuO_2 \bullet xH_2O$, which creates a large inner surface available for ion transport and was shown to be a favorable feature in the context of designing potentiometric sensors. Both materials were examined with SEM microscope, X-ray diffractometer, and contact angle microscope, and the results revealed that the hydrous form can be characterized as a porous structure with a smaller crystallite size and more hydrophobic properties contrary to the anhydrous form. Potentiometric and electrochemical tests carried out on designed $GCD/RuO_2/K^+$-ISM and $GCD/RuO_2 \bullet xH_2O/K^+$-ISM electrodes proved that the loose porous microstructure with chemically bounded water, which is characteristic for the hydrous form, ensures the high electrical capacitance of electrodes (up to 1.2 mF) with consequently more stable potential (with the potential drift of 0.0015 mV/h) and a faster response (of a few seconds).

Keywords: anhydrous and hydrous ruthenium dioxide; porous microstructure; high capacity; stable measuring signal

1. Introduction

Ruthenium dioxide can be considered as undeniably noteworthy amongst various electroactive materials previously used as mediation layers in solid-contact ion-selective electrodes (ISEs). One of the features characterizing solid-contact layer material is its electrical capacity, which is desired to be as high as possible to ensure the stable and fast potentiometric response of ISEs. As previously presented in [1,2], ruthenium dioxide allows achieving the highest capacitance of potassium-selective electrode with a polymeric membrane as a single-component layer (being conquered by hybrid layers consisting of two and more components). The success of ruthenium dioxide as a material for mediation layers in ion-selective electrodes lies in its porous structure and small grains providing high surface area and good charge-transfer characteristics [3,4].

In the literature, two main forms of the amorphous ruthenium dioxide, anhydrous (RuO_2) and hydrous form ($RuO_2 \bullet xH_2O$), are mentioned and described [5–9]; both of them were studied in the scope of the presented work and implemented as mediation layers in potassium-selective electrodes.

The two materials (anhydrous and hydrous ruthenium dioxide) consist of primary grains clustered together in larger secondary grains. The size, packing, and surface area of these derived grains depend on the water content in the oxide's structure [3]. Anhydrous RuO_2 is characterized by grater and densely packed grains in comparison to hydrous ruthenium dioxide and its loosely packed minute grains, which ensure its higher surface area.

The difference in the structure of both materials is also based on the difference in the type of pores. In the structure of ruthenium dioxide, two types of pores that participate in charge transfer processes can be found and characterized separately—mesopores, obtained from secondary particles, and micropores. Macropores can be compared to the highways for the ion transport, while low-conducting micropores can be related to the busy city roads. Due to the packed structure of anhydrous RuO_2, this type of oxide is low microporous, while in the loose morphology of hydrous RuO_2, substantial hydrated macropores can be found [10]. Therefore, in anhydrous ruthenium dioxide, the conduction processes are connected only to the occurring macropores, while in the hydrous form, both types of pores are effectively used for the conduction.

The slow and controlled ion conduction via micropores is responsible for the low grain-boundary charging and leading to the increase of the capacitance value of the hydrous form in contrast to the hydrous one, which uses only easy accessible mesopores [11].

Hence, the decrease of water in ruthenium oxide's structure that is chemically bound to the oxide's grains is attributed to the increase of the grain size and decrease in micropores content, which leads to the decrease in ionic capacitance.

The overall capacitance of various morphologies of RuO_2 is the totality of three partial capacitances C_{dl}, C_{ad}, and C_{irr}, which are named the electric double layer capacitance, the adsorption related charge, and the irreversible redox related charge, respectively [3]. Whilst the values of the non-faradaic electrical double layer capacitance of both hydrous and anhydrous forms of RuO_2 were found to be similar due to their comparable electrochemically active surface area, the values of C_{ad} and C_{irr} are varied and depend on the hydration degree. Hydrous oxide demonstrates larger C_{ad} values than its anhydrous form due to the large ion-diffusion occurring in hydrated nanoparticles yet smaller C_{irr} values, which can be attributed to the lack of grain-boundary charging. The measurements conducted in the scope of this work allowed comparing the total capacitance of both forms of ruthenium dioxide.

Another feature of ruthenium dioxide that has to be discussed is its charge-storage properties resulting from charge-transferred reactions between the oxide's layer surface and electrolyte. In both hydrous and anhydrous form, those processes are initiated by the electrons and protons introduced to the material's surface from electrolyte [2,12,13]. A detailed and suggested mechanism representing how RuO_2 works as an ion-to-electron transducer was presented in previous work [1].

Contrary to the hydrous form, anhydrous RuO_2 is reported to exhibit high electronic conductivity yet low proton mobility [14]. To increase the proton conductivity, structural water can be introduced into the rutile structure of ruthenium dioxide, as water film in-between grains allows both protons and electrons to be transported through the oxide's layer [3]. Moreover, the porous structure of $RuO_2 \bullet xH_2O$ ensures a shorter diffusion distance in contrast to the anhydrous form [11].

Taking into consideration all the mentioned features and differences between both studied materials, it can be concluded that hydrous ruthenium dioxide exhibits more favorable characteristics in the context of designing potentiometric sensors. Augustyn et al. [11] report that the high electrical capacitance and rapid faradaic reaction of $RuO_2 \bullet xH_2O$ are induced by its unique features, such as rapid electron transport caused by the metallic conductivity of RuO_2, rapid proton transport as a result of the presence of structural water, and the redox behavior of various $Ru-RuO_x$ states allowing for faradaic energy storage and a large surface area.

The differences between both forms of RuO_2 (hydrous and anhydrous) used for solid contact layers—those described in the Introduction and those presented in the Materials Characteristic section—have its reflection in the performance of ion-selective electrodes. Many different parameters such as the type of solvent, amount of oxide, hydration level, and layer thickness affect the electrical capacitance value. In order to help scientists who are concerned obtain similar results of capacitance for the ruthenium dioxide layer, in this article, we present a completely optimized procedure for the application of ruthenium oxide layers.

2. Experimental Section

2.1. Materials and Chemicals

Both studied forms of ruthenium dioxide—hydrous and anhydrous—were obtained commercially from Alfa Aesar (Haverhill, MA, USA) and Acros Organics (Fair Lawn, NJ, USA), respectively.

Potassium-selective membrane components—ionophore, lipophilic salt, plasticizer and polyvinyl chloride (PVC)—were purchased from Sigma Aldrich (St. Louis, MO, USA).

The organic solvents tetrahydrofuran (THF), dimethylformamide (DMF), and ethylene glycol (Gly) used for layer materials dispersion and membrane components dissolution were purchased from Sigma Aldrich.

Potassium Chloride KCl used to prepare standard potassium solutions was obtained from POCH (Gliwice, Poland). All materials were used as obtained without any further purification. For aqueous solutions preparation, distilled and deionized water was used.

2.2. Apparatus

Before being implemented onto electrodes, ruthenium dioxide (hydrous and anhydrous) was examined using a Scanning Electron Microscope (LEO 1530, Carl Zeiss, Wetzlar, Germany), contact angle microscope (Theta Lite with One Attension software by Biolin Scientific, Gothenburg, Sweden), differential scanning calorimeter (type DSC 2010, TA Instruments, New Castle, DE, USA), and X-ray diffractometer (X'Pert Pro with HighScore Plus software by PANalytical, Malvern, UK).

For the electrochemical tests carried on electrodes covered with studied layers of both types, the Autolab analyzer (Eco Chemie AUT32N.FRA2-AUTOLAB, ΩMetrohm, Herisau, Switzerland) analyzer) with NOVA software was used. All tests including cyclic voltammetry (CV), electrochemical impedance spectroscopy (EIS), and chronopotentiometry (Ch) were conducted in the 3-electrode cell consisting of a working electrode—a glassy carbon disc electrode covered with a RuO_2 layer, reference electrode, and auxiliary electrode—a glassy carbon rod. All tests conducted with the use of an Autolab analyzer were performed with the use of 10^{-1} M KCl acting as electrolyte.

Designed ion-selective electrodes, after being carefully studied using the mentioned electrochemical techniques, were tested toward examining their potentiometric performance with use of the 16-channel Lawson Labs potentiometer for this purpose. The potential of all studied electrodes was measured against the reference electrode Ag, AgCl/3M KCl (ΩMetrohm, 6.0733.100 type) and double-junction electrode (ΩMetrohm, 6.0729.100 type) for long-lasting stability tests and in the presence of a platinum rod acting as an auxiliary electrode.

2.3. Electrodes Preparation

Before being casted with the studied layers and Ion Selective Membrane (ISM), the Glassy Carbon Disc (GCD) electrodes' surface was prepared accordingly by polishing electrodes on alumina slurries (0.1 and 0.05 μm subsequently) and rinsing them with water and methanol.

In order to determine the difference between hydrous and anhydrous ruthenium dioxide in context of their influence on the ion-selective electrodes' properties, three groups of electrodes were studied. The first group consists of three coated-disc electrodes (GCD/K^+-ISM), which act as a control group providing the base results for modified RuO_2-contacted electrodes.

The other two groups include solid-contact electrodes (three per group)—one group with hydrous RuO_2 (GCD/$RuO_2 \bullet xH_2O$/K^+-ISM), and one with anhydrous RuO_2 (GCD/RuO_2/K^+-ISM) as the mediation layer.

The anhydrous RuO_2 layer was prepared by dispersing 10 mg of RuO_2 powder in 0.5 mL of DMF (which corresponds to the RuO_2 concentration of 20 mg/mL) with the use of ultrasonic washer. The hydrous oxide's layer was obtained analogously, and the concentration of $RuO_2 \bullet xH_2O$ was 7 mg/mL DMF.

Both layers were implemented onto electrodes by topping them with 15 µL of layer solution using the simple and fast drop-casting method. After casting, electrodes were left until complete solvent evaporation in the elevated temperature (approximately 70 °C). After removing the DMF, solid particles of RuO_2 remain on the electrode's surface and adhere physically to the electrode material. No additional chemicals or stabilizers are used to attach the layer to the GCD electrode.

Afterwards, six electrodes with dried layers (three per type) and three electrodes without any layer (control electrodes) were covered twice with 30 µL of membrane solution which consisted of potassium ionophore (I) 1.10% (w/w), lipophilic salt (KTpClPB) 0.25% (w/w), plasticizer (o-NPOE) 65.65% (w/w), and PVC 33.00% (w/w) dissolved in tetrahydrofuran. Casted electrodes were left to dry at room temperature.

All prepared electrodes (12 items) were conditioned in 0.01 M K^+ ion solution before being used for electrochemical measurements.

3. Results and Discussion

Before examining their suitability as mediation layers, both forms of ruthenium dioxide were subjected to tests in order to define their microstructure, wetting properties, crystallite size, and thermal behavior. All performed analysis enabled distinguishing and comparing the materials and helped reveal which of the material properties are crucial when applying them into potentiometric sensors.

3.1. Structure and Morphology Characterizations

SEM scans of both studied forms of ruthenium dioxide confirmed the differences mentioned in the introduction part. The grains of anhydrous ruthenium dioxide are significantly bigger in comparison comparing to those observed for the hydrous form. The hydrous RuO_2 material is formed with minute particles, while in anhydrous oxide, grains of various sizes and aspect ratios can be found.

As the grains observed on scans (Figure 1) are visibly agglomerated, to compare the size of single crystallites, the X-ray diffraction method was incorporated.

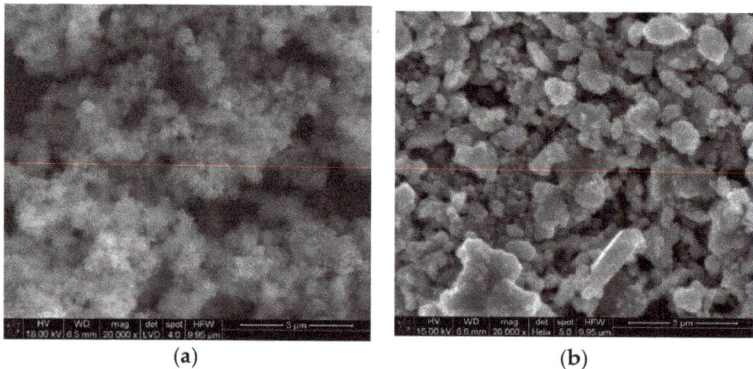

Figure 1. SEM scans of (a) hydrous RuO_2 and (b) anhydrous RuO_2.

The X-ray diffraction method was used for the structural characterization of both examined materials. The obtained X-ray diffraction patterns show differences between hydrous and anhydrous RuO_2, indicating that the water content in its rutile structure decides the peak width and peak intensity. As known from the literature [15,16], the peak intensity decreases and the peak width increases with the increasing water content. The increase of the peak width can be attributed to the decrease in the primary particle size, and the decrease in peak intensity suggests a decrease in the amount of crystalline RuO_2 and/or inferior crystallinity [17]. Based on the literature and obtained XRD patterns (Figure 2), it can be concluded that anhydrous RuO_2 is more crystalline and is characterized by a greater crystallite size contrary to the hydrous form. Figure 2a shows the amorphous halo with no diffraction peaks, which suggests that the hydrous form of oxide is less crystalline.

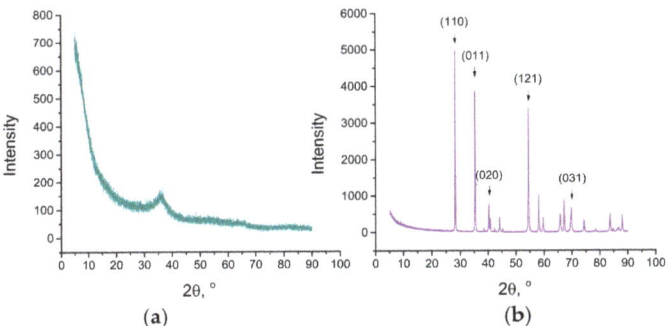

Figure 2. X-ray diffractogram of (**a**) hydrous RuO_2 and (**b**) anhydrous RuO_2.

The powder XRD pattern of the anhydrous RuO_2 (Figure 2b) contains the diffraction peaks at 28.1°, 35.1°, 40.1°, 54.3°, and 69.6°, which can be indexed to (110), (011), (020), (121), and (031) reflections, respectively [18]. The detected phase of ruthenium dioxide belongs to the 2/mnm space group in the tetragonal crystal system.

Contrary to the one of hydrous form, the diffractogram obtained for the anhydrous ruthenium dioxide contains well-defined, sharp reflections; hence, it was possible to determine the size of the single crystallites assembling this material.

Crystallite size (D) was calculated according the Scherrer equation $D = k\lambda/B\cos\theta$ [19] based on the wavelength (λ), shape factor (k), peak position (given by θ), and its full-width at half-maximum (B), with the use of HighScore Plus software. The size calculated from the data for the five most intensive peaks was of 69.8 ± 4.9 nm.

3.2. Hydration Structure and Wettability

Contact angle microscope tests allowed examining the degree of wetting for the studied materials. A water drop was released onto the surface of hydrous and anhydrous ruthenium dioxide, and the angle between the materials' surface and tangent to the formed water film was measured with the use of the microscope's software. The obtained microscope pictures (Figure 3) display that both types of oxide can be characterized as hydrophilic with the contact angle of 28° for the anhydrous form (Figure 3a) and 7° for the hydrous form (Figure 3b).

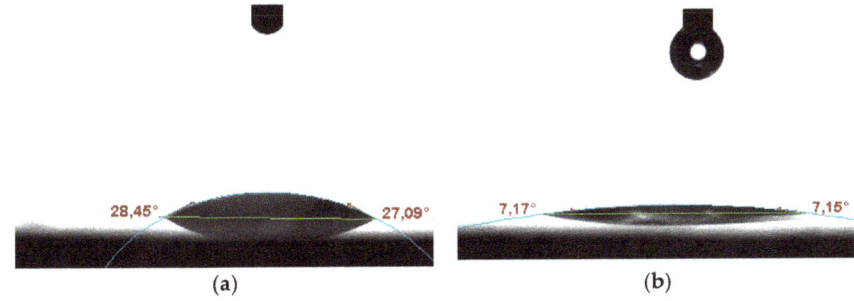

Figure 3. Contact angle microscope pictures of (**a**) anhydrous RuO_2, (**b**) hydrous RuO_2.

In order to characterize and compare the structures of hydrous and anhydrous ruthenium dioxide more thoroughly, DTA/TG (Differential Thermal Analysis/Thermogravimetry) plots are presented in Figure 4. First, 11 mg samples of RuO_2 and $RuO_2 \bullet H_2O$ powder were examined for the purpose of this measurement, and the temperature increased at a 10 °C/min rate in the range from 0 to 700 °C.

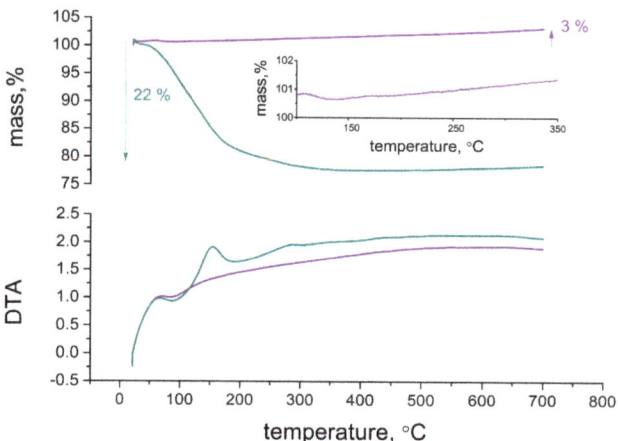

Figure 4. DTA/TG (Differential Thermal Analysis/Thermogravimetry) diagrams of anhydrous (violet curves) and hydrous (blue curves) RuO_2 (upper curves: TG analysis, bottom curves: DTA).

Thermogravimetry analysis was performed in order to estimate the weight alterations during the course of measurement. For hydrous ruthenium dioxide, the weight decrease occurs continuously from 0 to 300 °C, and the amount of weight loss in ruthenium oxide was measured to be approximately 22% of sample mass. As described in the literature [19], firstly, the mass loss is associated to the desorption of physically adsorbed water and secondly to the desorption of the chemically bounded water contained in the oxide's structure.

For anhydrous oxide, the weight firstly decreased, which could be attributed to the removal of impurities or water desorption from the surface of the oxide particles, and it eventually increased (3%) during the course of measurement. This phenomenon can be attributed to the oxidation process that transferred ruthenium IV into a higher oxidation state, which contributed to the weight gain of the sample mass [20].

The DTA results obtained for anhydrous RuO_2 depicted one peak at ca. 100 °C associated with the removal of physically adsorbed water on the surface of nanoparticles. An analogous peak was observed for hydrous RuO_2, which besides chemically bounded water, contains the same surface adsorbed water as the anhydrous one. On the DTA curve of hydrous ruthenium oxide nanoparticles, two more peaks were observed—at the temperatures near 150 °C and 300 °C, respectively. The first sharp peak, centered at ca. 150 °C, is attributed to the decomposition of chemically bounded water within nanoparticles. The second broad peak, at ca. 300 °C, is associated to the fusion and growth of ruthenium oxide nanoparticles to form larger crystals [19,21].

3.3. Electrochemical Capacitance Characterizations

Electrical parameters of solid-contact ISEs consisting of a mediation layer and ion-selective membrane can be considered as the result of values obtained for the membrane or layer itself. To learn about the properties of solid layers more thoroughly, layers can be studied after being implemented onto the glassy carbon electrode's surface using such techniques as cyclic voltammetry, electrochemical impedance spectroscopy, and chronopotentiometry. All the mentioned techniques allow determining the most significant parameter, which is electrical capacitance; hence, the results can be easily compared.

3.3.1. Cyclic Voltammetry Measurements

Firstly, electrical capacitance was determined using the cyclic voltammetry technique by cycling the potential of a working electrode and measuring its current response.

RuO_2-based layers were scanned with the speed of 0.1 V/s in the potential window between 0 and 0.3 V, and the CV curves for both studied materials were recorded and presented in Figure 5. The potential range was selected carefully so that no electrochemical reactions occur during the measurement and so the electrical capacitance value can be determined. The absolute current value observed in the middle of the potential window (that is, for the potential value of approximately 0.15 V) can be used for estimating the electrical capacity of the material, as divided by the scan rate (0.1 V/s), it gives the capacitance value. When comparing the curves recorded for the $RuO_2 \bullet xH_2O$ and RuO_2 layers, it can be observed that the first material is characterized by higher capacity than the other, as the current values are higher at the GCD/$RuO_2 \bullet xH_2O$ voltammogram. The obtained results are presented in Table 1.

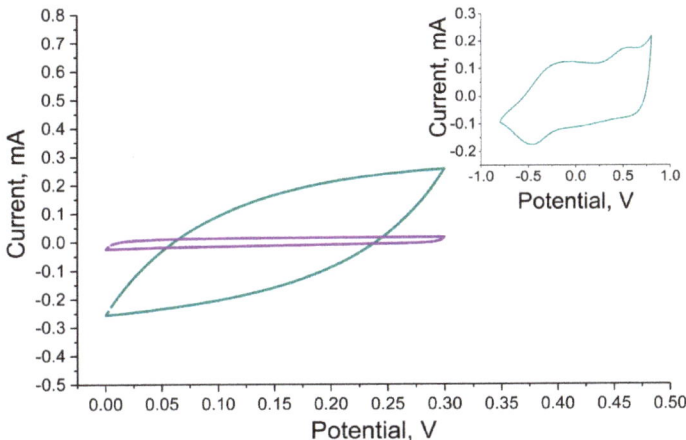

Figure 5. Comparison of electrochemical properties of anhydrous (violet curve) and hydrous (blue curve) RuO_2 tested with the use of cyclic voltammetry in the potential range from 0 to 0.3 V, inset: voltammogram of hydrous RuO_2 registered in a broader potential window from −1.0 to 1.0 V. Remaining cyclic voltammetry (CV) parameters: scan rate: 0.1 V/s, electrolyte: 0.01 M KCl.

Table 1. Electrical capacitance of layers (n = 3).

Group of Electrodes	Cyclic Voltammetry C ± SD [mF]	Electrochemical Impedance Spectroscopy C ± SD [mF]	Chronopotentiometry C ± SD [mF]
GCD/RuO_2	0.13 ± 0.01	0.27 ± 0.02	0.18 ± 0.01
GCD/$RuO_2 \bullet xH_2O$	2.5 ± 0.1	2.3 ± 0.1	2.6 ± 0.1

The inset to Figure 5 presents a voltammogram of hydrous ruthenium dioxide obtained using the same scanning speed yet a broader potential window (−1.0 to 1.0 V) in which it was possible to observe the peaks assigned to the ionic exchange processes and redox reactions as described by Majumdar et al. in [3].

3.3.2. Electrochemical Impedance Spectroscopy Measurements

The electrochemical impedance spectroscopy technique was also implemented onto the experiment to study the properties of the RuO_2-based layers. The Nyquist plots recorded for both studied materials were compared and presented in Figure 6a. Materials were examined in the frequencies range between 100 and 0.01 Hz with an amplitude of 0.01 V and set potential value of 0.15 V.

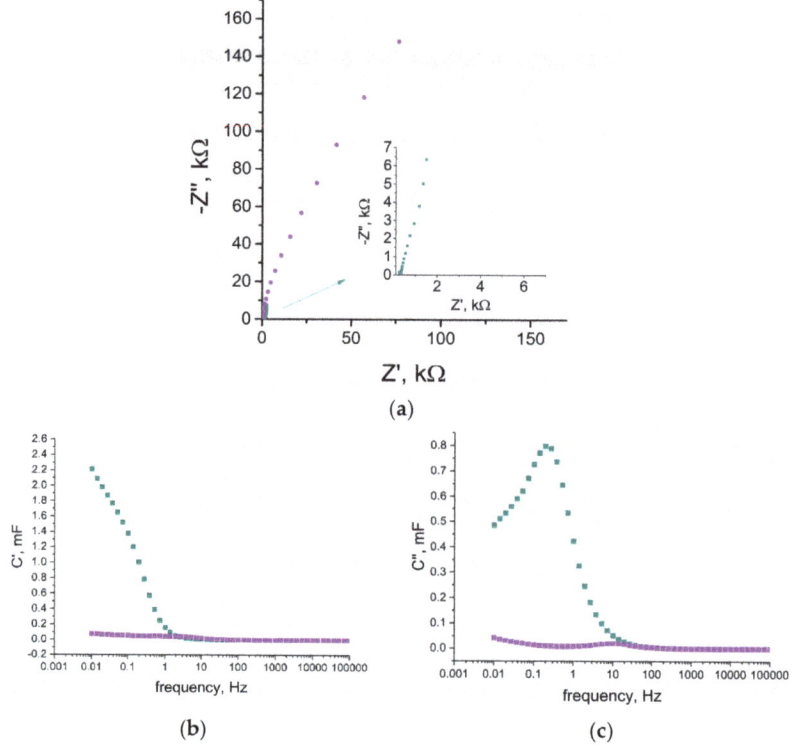

Figure 6. Comparison of electrochemical properties of anhydrous (violet curves) and hydrous (blue curves) RuO_2 tested with the use of electrochemical impedance spectroscopy in the frequency range of 100 kHz and 0.01 Hz with an amplitude of 0.01 V and set potential value of 0.15 V in 0.01 M KCl as electrolyte, (**a**) electrochemical impedance spectroscopy (EIS) curves of both tested layers, inset: closer look at hydrous RuO_2, (**b**) real part of capacitance (C′) change vs. frequency, (**c**) imaginary part of capacitance (C″) change vs. frequency.

For low frequencies, the capacitance (C) value can be calculated using the $C = 1/(2\pi Zf)$ equation, where f stands for the frequency (the lowest examined: f = 0.01 Hz) and Z stands for the imaginary part of the impedance. The obtained capacitance value was juxtaposed with the results from the other techniques in Table 1.

Using the real part of the capacitance versus frequency plot (Figure 6b), it can be seen that with the decrease in frequency, C′ increases sharply for the hydrous RuO_2 material, contrary to the flat dependency observed for anhydrous RuO_2. The capacitance reaches around 2.2 mF for $RuO_2 \bullet xH_2O$ and 0.15 mF for RuO_2, yet the whole capacitance is not reached (the value still increases with the decreasing frequency). This phenomena occurs because ions from the electrolyte have not reached the whole electrode's material porosity.

Figure 6c presents the change of C″ versus frequency. The imaginary part of the capacitance C′ goes through a maximum at a frequency f_0, with a time constant defined as $t_0 = 1/f_0$ [22]. This time constant is described as a dielectric relaxation time that characterizes the whole system [23]. The relaxation time for hydrous RuO_2 is 5 s and for anhydrous, it is 0.07 s.

3.3.3. Chronopotentiometry

Electrical parameters, which can be determined with chronopotentiometry, such as capacitance and resistance, can be used to characterize both electroactive layers and electrodes with polymer membranes including those layers. To examine the layer itself, the electroactive material is dropped in form of the solution on the surface of the glassy carbon disc electrode and left until the solvent evaporation. Solid-contact and coated-disc electrodes are studied as obtained (according to the preparation procedure described in the work) without any modifications.

Both layers and layers with membranes were examined using the same technique, and the results are presented in Figures 7–9, respectively.

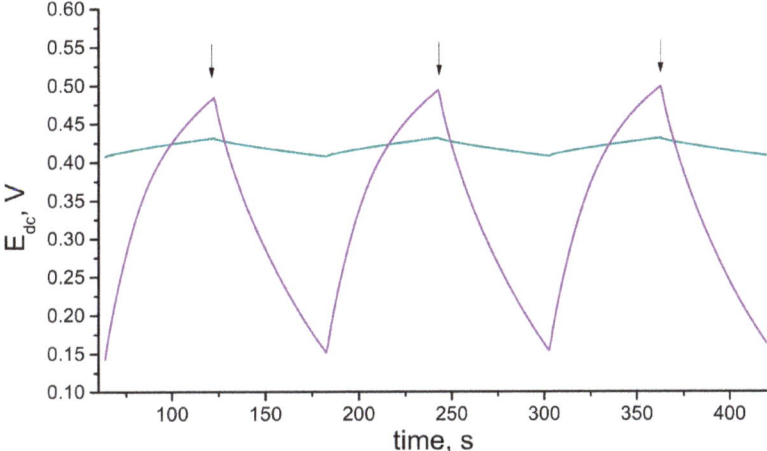

Figure 7. Comparison of electrochemical properties of anhydrous (violet) and hydrous (blue) RuO_2 tested with the use of chronopotentiometry. Arrows point the moment of the current sign change.

The electrical capacity of RuO_2-based layers and further solid-contact electrodes (C) was determined using the chronopotentiometry technique. The principle of this electrochemical method underlies recording the potential of an electrode (E_{dc}) while the current (I) is forced to flow through the 3-electrode cell (which consists of a working GCD electrode, reference electrode, and auxiliary electrode). The method was programmed for the time (t) of 60 s of +1 nA current flow followed by 60 s of −1 nA current flow. The program covered six steps and lasted 360 s in total. After every 60 s, the current sign changed, the potential jump (ΔE_{dc}) is observed, and the potential is recorded for the following 60 s (with the opposite current sign). This method allows calculating the capacitance value using the mentioned parameters and the $\Delta E_{dc}/\Delta t = I/C$ equation.

As the capacitance value of electrodes determines their analytical and electrical performance, this parameter was used to evaluate and compare layers of different types (form), thicknesses, or other varied factors. Therefore, the chronopotentiometry method was applied to select the best type amongst the studied materials, that is, the one of the highest capacitance value. The aim was to select the most favorable solvent for RuO_2 particles and the most optimum volume of layer solution to be casted on the electrode, and eventually, to compare the electrical capacity of hydrous and anhydrous form of ruthenium dioxide. Tested solvents included THF, DMF, and ethylene glycol (Gly), and all of them were used to prepare solutions of hydrous and anhydrous forms of oxide. The capacitance values obtained for each solvent and layer volume were presented in columns and compared in Figure 8.

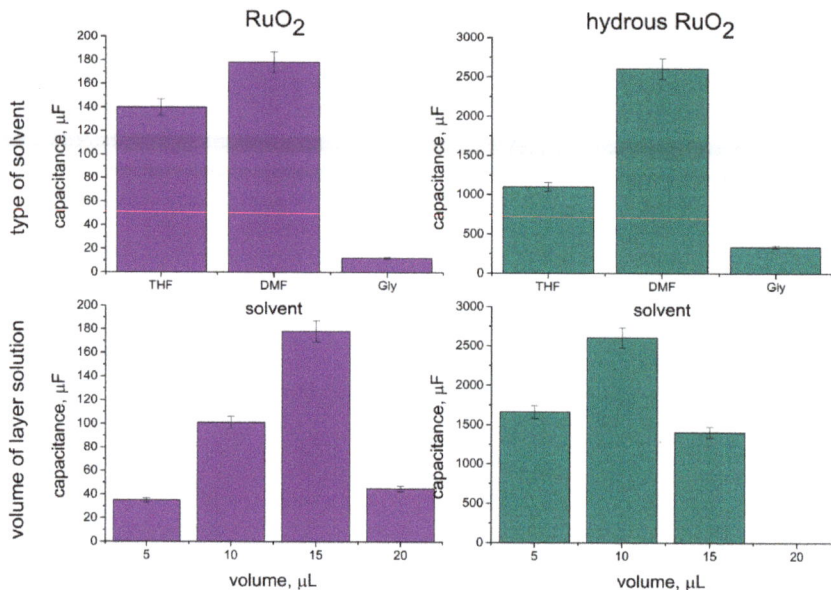

Figure 8. Capacitance values obtained for anhydrous (violet) and hydrous (blue) ruthenium dioxide dispersed in various solvents and applied in different amounts onto the electrode's surface.

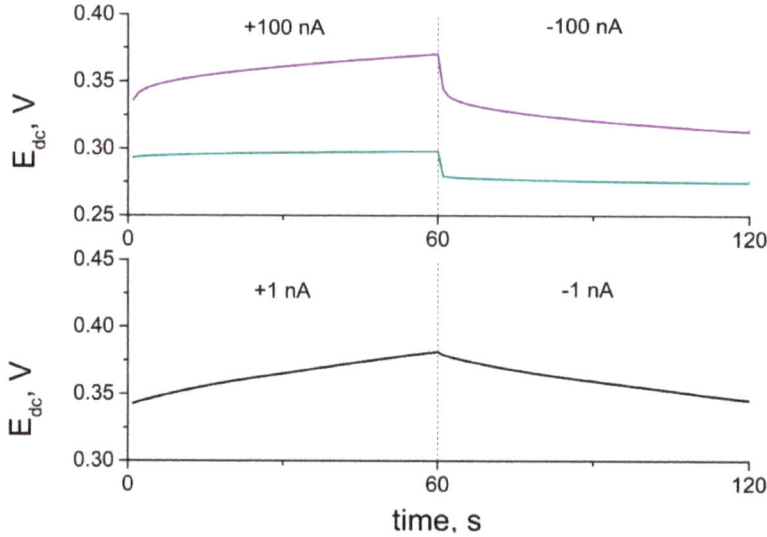

Figure 9. Part of chronopotentiograms recorded for solid contact electrodes: GCD/RuO$_2$•xH$_2$O/K$^+$-ISM electrode (blue curve), GCD/RuO$_2$/K$^+$-ISM electrode (violet curve) and coated disc GCD/K$^+$-ISM electrode (black curve).

First, 10 µL of each solution was casted onto electrodes, and the capacitance values were as follows: 140 µF, 12 µF, and 101 µF for the THF-RuO$_2$, Gly-RuO$_2$, and DMF-RuO$_2$ layers, respectively and 1098 µF, 336 µF, and 2602 µF for the THF-RuO$_2$•xH$_2$O, Gly-RuO$_2$•xHO, and DMF-RuO$_2$•xH$_2$O layers.

As presented, the highest capacitance value was obtained for both DMF-based studied layers, which allowed indicating this solvent as the most favorable in the context of designing solid-contact electrodes. Between the hydrous and anhydrous form of RuO$_2$, there is a significant difference in the electrical capacity of both materials, and the hydrous form turned out to exhibit a much higher capacitance value.

After selecting the right solvent, layers of different thicknesses were examined. DMF-based RuO$_2$ and RuO$_2$•xH$_2$O solutions were dropped onto the electrodes surface in the following amounts: 5 µL, 10 µL, 15 µL, and 20 µL; the results were as presented: 35 µF, 101 µF, 178 µF, and 45 µF for anhydrous form and 1660 µF, 2602 µF, and 1404 µF for hydrous form. The layer of 20 µL was not examined in the case of hydrous form as after placing the electrode into the K$^+$ ion solution, the RuO$_2$•xH$_2$O particles were detached from the electrode's surface. The selected volume for RuO$_2$ DMF-based layer solution was of l5 µL and for the hydrous form, it was of 10 µL.

The values of capacitance (for the optimized DMF-based layers) obtained using the chronopotentiometry technique were presented and compared with the other two applied electrochemical techniques (CV and EIS) in Table 1. The values of electrical capacitance obtained from all implemented techniques were in good agreement for both studied types of layers. The approximate capacitance value for the anhydrous RuO$_2$ layer was of 0.2 mF, which is more than 10 times lower in comparison to the value achieved by the hydrous layer (approximately 2.5 mF). Those differences confirmed that the morphological and structural properties of the layers translate into their electrochemical characteristics.

When comparing the values presented in Table 1 (for layers) and those from the last column in Table 2 (for designed ion-selective electrodes), it can be seen that the properties of the layers strongly determine the properties of the designed electrodes. The differences observed between the RuO$_2$ and RuO$_2$•xH$_2$O layers can be also recognized in between GCD/RuO$_2$•xH$_2$O/K$^+$-ISM and GCD/RuO$_2$/K$^+$-ISM electrodes. Implementing the RuO$_2$•xH$_2$O layer of high electrical capacity allowed obtaining electrodes characterized by considerably high capacitance, while the capacitance of the GCD/RuO$_2$/K$^+$-ISM electrodes turned out to be more than 10 times smaller.

Table 2. Electrical parameters of studied electrodes (n = 3). GCD: glassy carbon disc.

Group of Electrodes	Resistance ± SD [kΩ]	Potential Drift ± SD [µV/s]	Capacitance ± SD [µF]
GCD/K$^+$-ISM	1022 ± 21	598 ± 20	1.67 ± 0.06
GCD/RuO$_2$/K$^+$-ISM	279.5 ± 0.5	532.8 ± 0.6	188 ± 2
GCD/RuO$_2$•xH$_2$O/K$^+$-ISM	94.64 ± 0.07	81.1 ± 0.9	1233 ± 14

3.4. Electrical Parameters of Electrodes

3.4.1. Chronopotentiometry

With the use of the same technique (chronopotentiometry) and the same parameters that need to be recorded to obtain the capacitance value, another two important electrical parameters can be determined: resistance and potential drift.

While it is desired for the electrical capacitance to be of the highest value, the resistance and potential drift are expected to be as low as possible. The resistance value can be calculated using potential jump (ΔE_{dc}) and current (I) values with the use of the $R_{total} = \Delta E_{dc}/2I$ equation. Potential drift is defined as the potential change divided by the time change, which makes the simple equation: $\Delta E_{dc}/\Delta t$.

All the electrical parameters' values were presented in the table as averaged values calculated from results obtained for 3 electrodes from each group along with their standard deviations.

As presented, electrodes with a layer of hydrous RuO$_2$ exhibited the highest electrical capacity, and consequently, the lowest resistance and potential drift of all tested groups. Electrodes from the GCD/RuO$_2$/K+-ISM group, which are based on anhydrous oxide, presented considerably lower capacitance value and higher values of other electrical parameters. As mentioned in the Introduction, one of the differences between those two materials is the occurrence of structural water in RuO$_2$•xH$_2$O, which creates a large inner surface available for ion transport [11]. The total capacitance value of the hydrous form of oxide compared to the anhydrous one turned out to be much higher; therefore, the hydrous layer is more favorable in context of designing ion-selective electrodes.

The results obtained using the Autolab analyzer allowed characterizing layers before their implementation onto electrodes (Figure 7) and further characterizing electrodes (Figure 9) before using them in potentiometric tests and experiments.

3.4.2. Electrical Impedance Spectroscopy

Here, Electrical Impedance Spectroscopy was implemented again in order to determine the electrical parameters of RuO$_2$-based electrodes covered with an ion-selective polymeric membrane. Impedance data were collected in the frequency range from 100 kHz to 0.01 Hz using an amplitude of 10 mV superimposed onto open-circuit potential (OCP). The EIS plots were presented in Figure 10 together with the equivalent circuit. The solid line represents the EIS data fitted to obtained EIS results using NOVA 2.1.4 software. The diameter of a semicircle in the high-frequency region represents the bulk membrane resistance R_b, which for the GCD/RuO$_2$•xH$_2$O/K$^+$-ISM electrode is 105 kΩ, and for the GCD/RuO$_2$/K$^+$-ISM electrode, it is 269 kΩ. The bulk resistance R_b results from the resistance of the ion-selective membrane and the resistance of the ISM inside the porous structure of the mediation layer; therefore, the resistance of the electrode with more porous material such as hydrous RuO$_2$ is lower. In the low-frequency region, the EIS curve results from double-layer capacitance (CPE$_{dl}$) and charge-transfer resistance (R_{ct}), which provides the information about the ease of electron/ion transfer at the electrode interface. As presented in Table 3, an electrode with hydrous RuO$_2$ as the mediation layer is characterized with lower resistance and higher capacitance (R_{ct} = 24 kΩ, CPE = 1106$^{(0.443)}$ μS$^{(N)}$ for the GCD/RuO$_2$•xH$_2$O/K$^+$-ISM electrode) in contrast to the GCD/RuO$_2$/K$^+$-ISM electrode, for which R_{ct} = 128 kΩ and CPE = 185$^{(0.702)}$ μS$^{(N)}$. Therefore, it can be concluded that hydrous RuO$_2$ allows fast charge transfer processes and guarantees a stable potential response of designed ISEs.

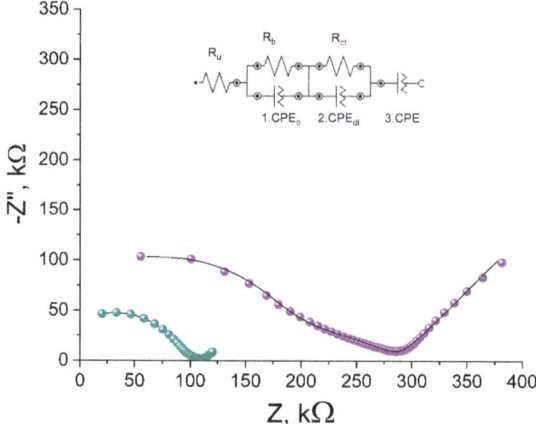

Figure 10. Impedance spectrum of GCD/RuO$_2$•xH$_2$O/K$^+$-ISM (blue) and GCD/RuO$_2$/K$^+$-ISM (violet) electrode in 0.01 M KCl solution and equivalent electrical circuits. Frequency range: 100 kHz–0.01 Hz. Equivalent circuits are shown as insets (solid lines represent data fits).

Table 3. Electrical parameters of studied electrodes.

Electrode Type	R_u (kΩ)	R_b (kΩ)	CPE_b (pS)$^{(N)}$	R_{ct} (kΩ)	CPE_{dl} (nS)$^{(N)}$	CPE (μS)$^{(N)}$
GCD/RuO$_2$•xH$_2$O/K$^+$-ISM	−32	105	34.2$^{(0.93)}$	24	968$^{(0.294)}$	1106$^{(0.443)}$
GCD/RuO$_2$/K$^+$-ISM	−48	269	26.1$^{(0.87)}$	128	117$^{(0.409)}$	185$^{(0.702)}$

The bulk resistance and double-layer capacitance obtained with the use of EIS are comparable to the electrical parameters (resistance and capacitance) of studied electrodes evaluated using the Chronopotentiometry method.

3.5. Potentiometric Tests

Designed electrodes after being tested with the use of various electrochemical techniques, where the current value was a significant value, were implemented into the traditional potentiometric measurement. All studied groups, including two groups of solid-contact electrodes (GCD/RuO$_2$/K$^+$-ISM and GCD/RuO$_2$•xH$_2$O/K$^+$-ISM) and a group of coated-disc electrodes (GCD/K$^+$-ISM), were connected with potentiometer and placed into the measuring vessel together with reference and auxiliary electrodes in order to study their behavior in the conditions of typical potentiometric measurement (with no current presence).

Since the desired application of the presented sensors is the measurement of potassium content in the water samples, firstly, the sensors must have been characterized and their parameters strictly determined. Electrodes were examined with the use of the potentiometry method and defined by indicating the slope of the calibration curve and the standard potential.

The calibration curves are presented in Figure 11. As shown, the detection limit for the GCD/RuO$_2$•xH$_2$O/K$^+$-ISM group is the lowest and reaches 10^{-6} M K$^+$. Implementing anhydrous RuO$_2$ into electrodes allowed obtaining a detection limit of $5\bullet10^{-5.5}$ M K$^+$. Both forms of ruthenium dioxide were found to improve the linear range of sensors, as for the coated-disc electrodes, the detection limit was only 10^{-5} M.

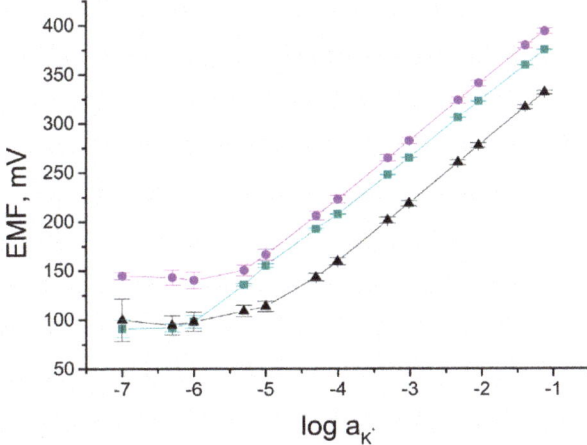

Figure 11. Calibration curves recorded for GCD/RuO$_2$•xH$_2$O/K$^+$-ISM (■), GCD/RuO$_2$/K$^+$-ISM (•), and GCD/K$^+$-ISM (▲) electrode.

Based on the parameters of curves recorded in the function of Electromotive Force and logarithm of activity of K$^+$, the equations were fixed for each group:

$$E = 57.37 \log a_{K^+} + 439 \text{ [mV] for CGD/RuO}_2 \bullet xH_2O/K^+\text{-ISM group} \quad (1)$$

$$E = 59.44 \log a_{K^+} + 462 \text{ [mV] for CGD/RuO}_2/K^+\text{-ISM group} \quad (2)$$

$$E = 57.01 \log a_{K^+} + 394 \text{ [mV] for GCD/K}^+\text{-ISM group.} \quad (3)$$

With the use of designed sensors and presented equivalences, it is possible to estimate the potassium content based on the measured potential value.

Comparing the convergence of results within a group of electrodes, it was possible to fix the average parameters for each group and determine the repeatability of the potential response. Based on the error bars presented in Figure 10, which correspond to the values of standard deviations calculated for the studied electrodes, it can be seen that the best compatibility between results was observed for the GCD/RuO$_2 \bullet$xH$_2$O/K$^+$-ISM group.

What should be emphasized here is that amongst the solid-contact potassium selective electrodes presented in the literature (a select few are presented in Table 4), RuO$_2 \bullet$xH$_2$O-based electrodes exhibit satisfying analytical parameters such as theoretical Nernstian response in the wide K$^+$ ions concentration range, moderate potential drift, and considerably high electrical capacitance. Therefore, it can be concluded that hydrous ruthenium dioxide applied alone and without any modifications allows obtaining the ion-selective electrode of parameters that are comparable to the best potentiometric methods described in the literature.

Table 4. Electrical and potentiometric parameters compared for GCD/SC/K$^+$-ISM electrodes with various materials applied as solid-contact (SC) layers: TCNQ—Tetracyanoquinodimethane; GR—Graphene; PEDOT(CNT)—Poly (3,4-ethylenedioxythiophene)–Carbon Nanotubes; MoO$_2$—Molybdenum Dioxide; CIM—Colloid-Imprinted Mesoporous Carbon; CB-GR-FP—Carbon Black–Graphene–Fluorinated acrylic copolymer.

Electrode	Capacitance Value [µF]	Potential Drift [µV/s]	Linear Range [M]	Slope [mV/dec]	Reference
GCD/TCNQ/K$^+$-ISM	132	12	10^{-1}–$10^{-6.5}$	58.68	[24]
GCD/GR/K$^+$-ISM	91	12	10^{-1}–$10^{-4.5}$	59.2	[25]
GCD/PEDOT(CNT)/K$^+$-ISM	83	12	10^{-1}–10^{-6}	57.7	[26]
GCD/MoO$_2$/K$^+$-ISM	86	11.67	10^{-3}–10^{-5}	55.0	[27]
GCD/CIM/K$^+$-ISM	1000	1	10^{-1}–$10^{-5.2}$	59.5	[28]
GCD/CB-GR-FP/K$^+$-ISM	1471	0.68	10^{-1}–$10^{-6.5}$	59.10	[29]
GCD/RuO$_2$/K$^+$-ISM	188	533	10^{-1}–$10^{-5.5}$	59.44	this work
GCD/RuO$_2 \bullet$xH$_2$O/K$^+$-ISM	1233	81	10^{-1}–10^{-6}	57.37	this work

3.6. Stability of Response

In the addition to the presented chronopotentiometric results was the potential drift, which was calculated based on the value of the potentiometric response and forced current flow; herein, we present the stability of the designed sensors in terms of the traditional potentiometric measurement. The potential was recorded over the time of 21 h, and the potential drift was estimated based on the obtained results.

The stability of electrodes given by the potential/time ratio was as follows: 0.0015 mV/h for the GCD/RuO$_2 \bullet$xH$_2$O/K$^+$-ISM electrode, 0.042 mV/h for the GCD/RuO$_2$/K$^+$-ISM electrode, and 0.48 mV/h for the coated-disc electrode. The results obtained at this stage are consistent with those of chronopotentiometry, as the electrical capacitance, which decides the electrodes' ability to exhibit a stable potentiometric response, was of the highest value for sensors with an RuO$_2 \bullet$xH$_2$O mediation layer.

Figure 12 presents the potential response curves, which reflect the response time of the fabricated ISEs. The exemplary potential stability of the GCD/RuO$_2$•xH$_2$O/K$^+$-ISM and GCD/RuO$_2$/K$^+$-ISM electrodes measured in a 0.01 M KCl solution during the first 10 min of measurement is presented in Figure 12a,b, respectively.

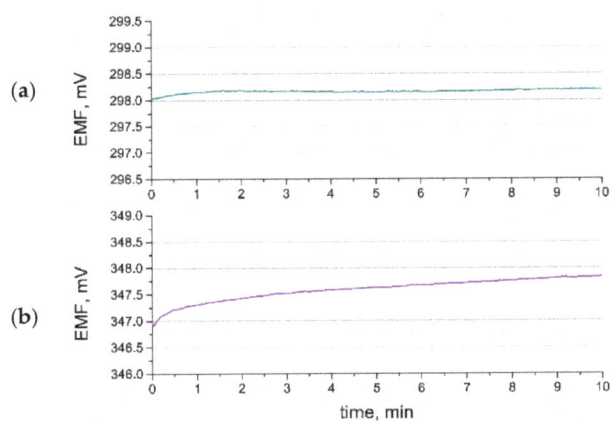

Figure 12. Potentiometric response of (**a**) GCD/RuO$_2$•xH$_2$O/K$^+$-ISM electrode and (**b**) GCD/RuO$_2$/K$^+$-ISM electrode during first 10 min of measurement.

According to the IUPAC (International Union of Pure and Applied Chemistry) nomenclature for ion-selective electrodes, the time of response is equal to the time period between contacting the analyte and reaching 95% of the equilibrium potential value. As shown, the time of response for a hydrous RuO$_2$-based electrode is as short as a few seconds, and the stable potential value is reached almost immediately. For the RuO$_2$-based electrode, the equilibrium potential is reached after a few minutes (a 95% potential value is reached immediately), and the potential stability is worse in comparison with the GCD/RuO$_2$•xH$_2$O/K$^+$-ISM electrode.

3.7. Water Layer Test

The water layer test was conducted to examine the water uptake of studied electrodes during the course of potentiometric measurement. Water from measured samples and conditioning solution tends to penetrate the membrane and form the thin film between the polymeric membrane and electronic conductor. This process may cause the potential drift of the electrodes' response and result in deterioration of the membrane's adherence; therefore, it is desired to eliminate the water layer. Water uptake can be limited by introducing the mediation layers that will prevent the accumulation of water. The ability of the studied ruthenium dioxide layers to eliminate the water layer was evaluated during a water layer test when exchanging primary ion (potassium) solution into sodium ion solution. Afterwards, the electrodes were placed into 10^{-2} M KCl solution for about 10 h, KCl was exchanged into 10^{-2} M NaCl solution to examine the potential drift, and after 5 h, it was exchanged back to the primary ion solution to examine the stability of potentiometric response. The potentiometric response was recorded with time, and the results are presented in Figure 13.

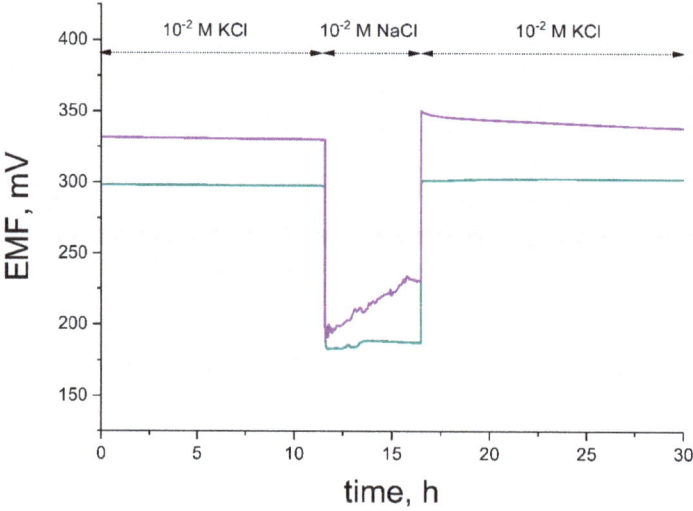

Figure 13. Water layer test of GCD/RuO$_2$•xH$_2$O/K$^+$-ISM (blue curve) and GCD/RuO$_2$/K$^+$-ISM (violet curve) electrode.

As shown, the response of the hydrous RuO$_2$-based electrode was stable after contacting NaCl solution and the characteristic potential drift was not observed, contrary to the response of the electrode with an anhydrous layer. After being placed back into primary ion solution, the GCD/RuO$_2$/K$^+$-ISM electrode exhibited a substantial drift of potentiometric response, and the original equilibrium for K$^+$ ions was established again after a few hours. The difference in behavior of two studied electrodes can be explained by the participation of structural water in the ion-exchange processes.

4. Conclusions

The differences between the hydrous and anhydrous form of ruthenium dioxide can be noticed at the very beginning of material characterization—the loose and porous microstructure of hydrous form with small crystallites and chemically bounded water is in opposition to the densely packed, macroporous structure of the anhydrous form. All the tested features of the studied materials—microstructure, wettability, and crystallite size—contribute significantly to the properties of materials as mediation layers in ion-selective electrodes. The small size of ruthenium dioxide particles ensured a high surface area, and the presence of the water in between grains provided extra pathways for ion transport; therefore, hydrous ruthenium dioxide turned out to be a more favorable material in the context of designing ion-selective electrodes.

This claim is supported by the results from chronopotentiometry, which revealed the high value of the electric capacitance and potentiometry. Potentiometric tests proved that the analytical performance of RuO$_2$-based electrodes is considerably better compared to that of a non-modified coated-disc electrode. Between the hydrous and anhydrous ruthenium dioxide layers, we selected RuO$_2$•xH$_2$O-based electrodes as the most robust, with their wider linear range (up to 10^{-6} M K$^+$) and stable potentiometric response (with the potential drift for this type of electrode of 0.0015 mV/h).

Author Contributions: Conceptualization, N.L., B.P.-B.; methodology, B.P.-B.; validation, B.P.-B., N.L. and R.P.; formal analysis, B.P.-B., N.L.; investigation, N.L.; resources, B.P.-B., R.P.; data curation, N.L.; writing—original draft preparation, N.L.; writing—review and editing, B.P.-B., N.L.; visualization, N.L.; supervision, B.P.-B.; project administration, B.P.-B.; funding acquisition, B.P.-B. All authors have read and agreed to the published version of the manuscript.

Funding: The publication is financed from the subsidy no 16.16.160.557 of the Polish Ministry of Science and Education.

Conflicts of Interest: The authors declare no conflict of interest.

References

1. Lenar, N.; Paczosa-Bator, B.; Piech, R. Ruthenium Dioxide as High-Capacitance Solid-Contact Layer in K$^+$—Selective Electrodes Based on Polymer Membrane. *J. Electrochem. Soc.* **2019**, *166*, 1470–1476. [CrossRef]
2. Lenar, N.; Paczosa-Bator, B.; Piech, R. Ruthenium dioxide nanoparticles as a high-capacity transducer in solid-contact polymer membrane-based pH-selective electrodes. *Microchim. Acta* **2019**, *186*, 1–11. [CrossRef] [PubMed]
3. Majumdar, D.; Maiyalagan, T.; Jiang, Z. Recent Progress in Ruthenium Oxide-Based Composites for Supercapacitor Applications. *ChemElectroChem* **2019**, *6*, 4343–4372. [CrossRef]
4. Zheng, J.P.; Cygan, P.J.; Jow, T.R. Hydrous Ruthenium Oxide as an Electrode Material for Electrochemical Capacitors. *J. Electrochem. Soc.* **1995**, *142*, 2699–2703. [CrossRef]
5. Majumdar, D. An Overview on Ruthenium Oxide Composites—Challenging Material for Energy Storage Applications. *Mater. Sci. Res. India* **2018**, *15*, 30–40. [CrossRef]
6. Deng, L.; Wang, J.; Zhu, G.; Kang, L.; Hao, Z. RuO2/graphene hybrid material for high performance electrochemical capacitor. *J. Power Sources* **2014**, *248*, 407–415. [CrossRef]
7. Yu, G.; Xie, X.; Pan, L.; Bao, Z.; Cui, Y. Hybrid nanostructured materials for high-performance electrochemical capacitors. *Nano Energy* **2012**, *2*, 213–234. [CrossRef]
8. Li, Q.; Zheng, S.; Xu, Y.; Xue, H.; Pang, H. Ruthenium based materials as electrode materials for supercapacitors. *Chem. Eng. J.* **2018**, *333*, 505–518. [CrossRef]
9. Shi, F.; Li, L.; Wang, X.; Gu, C.; Tu, J. Metal oxide/hydroxide-based materials for supercapacitors. *RSC Adv.* **2014**, *4*, 41910–41921. [CrossRef]
10. Makino, S.; Ban, T.; Sugimoto, W. Towards Implantable Bio-Supercapacitors: Pseudocapacitance of Ruthenium Oxide Nanoparticles and Nanosheets in Acids, Buffered Solutions, and Bioelectrolytes. *J. Electrochem. Soc.* **2015**, *162*, A5001–A5006. [CrossRef]
11. Augustyn, V.; Simon, P.; Dunn, B. Pseudocapacitive oxide materials for high-rate electrochemical energy storage. *Energy Environ. Sci.* **2014**, *7*, 1597–1614. [CrossRef]
12. Fog, A.; Buck, R.P. Electronic semiconducting oxides as pH sensors. *Sens. Actuators* **1984**, *5*, 137–146. [CrossRef]
13. Wen, S.; Lee, J.; Yeo, I.; Park, J.; Mho, S. The role of cations of the electrolyte for the pseudocapacitive behavior of metal oxide electrodes. *Electrochim. Acta* **2004**, *50*, 849–855. [CrossRef]
14. Dmowski, W.; Egami, T.; Swider-lyons, K.E.; Love, C.T.; Rolison, D.R. Local Atomic Structure and Conduction Mechanism of Nanocrystalline Hydrous RuO$_2$ from X-ray Scattering. *J. Phys. Chem. B* **2002**, *106*, 12677–12683. [CrossRef]
15. Lee, J.; Arif, S.; Shah, S.; Yoo, P.J.; Lim, B. Hydrous RuO$_2$ nanoparticles as highly active electrocatalysts for hydrogen evolution reaction. *Chem. Phys. Lett.* **2017**, *673*, 89–92. [CrossRef]
16. Sugimoto, W.; Iwata, H.; Yokoshima, K.; Murakami, Y. Proton and Electron Conductivity in Hydrous Ruthenium Oxides Evaluated by Electrochemical Impedance Spectroscopy: The Origin of Large Capacitance. *J. Phys. Chem. B* **2005**, *109*, 7330–7338. [CrossRef]
17. Bunaciu, A.A.; Aboul-enein, H.Y. X-ray Diffraction: Instrumentation and Applications Critical Reviews in Analytical Chemistry X-Ray Diffraction: Instrumentation and Applications. *Crit. Rev. Anal. Chem.* **2015**, *45*, 289–299. [CrossRef]
18. Cotton, F.A.; Mague, J.T. The Crystal and Molecular Structure of Tetragonal Ruthenium Dioxide. *Inorg. Chem.* **1966**, *5*, 317–318. [CrossRef]
19. Kim, J.; Kim, K.; Park, S.; Kim, K. Microwave-polyol synthesis of nanocrystalline ruthenium oxide nanoparticles on carbon nanotubes for electrochemical capacitors. *Electrochim. Acta* **2010**, *55*, 8056–8061. [CrossRef]
20. Audichon, T.; Napporn, T.W.; Cana, C.; Kokoh, K.B. IrO$_2$ Coated on RuO$_2$ as Efficient and Stable Electroactive Nanocatalysts for Electrochemical Water Splitting. *J. Phys. Chem. C* **2016**, *120*, 2562–2573. [CrossRef]

21. Kim, I.; Kim, K. Ruthenium Oxide Thin Film Electrodes Prepared by Electrostatic Spray Deposition and Their Charge Storage Mechanism. *J. Electrochem. Soc.* **2004**, *151*, E7–E13. [CrossRef]
22. Taberna, P.L.; Simon, P.; Fauvarque, J.F. Electrochemical Characteristics and Impedance Spectroscopy Studies of Carbon-Carbon Supercapacitors. *J. Electrochem. Soc.* **2003**, *150*, A292–A300. [CrossRef]
23. Cole, K.S.; Cole, R.H. Dispersion and Absorption in Dielectrics I. Alternating Current Characteristics. *J. Chem. Phys.* **2004**, *9*, 341–351. [CrossRef]
24. Paczosa-Bator, B.; Pięk, M.; Piech, R. Application of Nanostructured TCNQ to Potentiometric Ion-Selective K^+ and Na^+ Electrodes. *Anal. Chem.* **2015**, *87*, 1718–1725. [CrossRef]
25. Li, F.; Ye, J.; Zhou, M.; Gan, S.; Zhang, Q.; Han, D.; Niu, L. All-solid-state potassium-selective electrode using graphene as the solid contact. *Analyst* **2012**, *137*, 618–623. [CrossRef]
26. Mousavi, Z.; Bobacka, J.; Lewenstam, A.; Ivaska, A. Poly(3,4-ethylenedioxythiophene) (PEDOT) doped with carbon nanotubes as ion-to-electron transducer in polymer membrane-based potassium ion-selective electrodes. *J. Electroanal. Chem.* **2009**, *633*, 246–252. [CrossRef]
27. Zeng, X.; Qin, W. A solid-contact potassium-selective electrode with MoO_2 microspheres as ion-to-electron transducer. *Anal. Chim. Acta* **2017**, *982*, 72–77. [CrossRef]
28. Hu, J.; Zou, X.U.; Stein, A.; Bühlmann, P. Ion-selective electrodes with colloid-imprinted mesoporous carbon as solid contact. *Anal. Chem.* **2014**, *86*, 7111–7118. [CrossRef]
29. Paczosa-Bator, B. Ion-selective electrodes with superhydrophobic polymer/carbon nanocomposites as solid contact. *Carbon N. Y.* **2015**, *95*, 879–887. [CrossRef]

© 2020 by the authors. Licensee MDPI, Basel, Switzerland. This article is an open access article distributed under the terms and conditions of the Creative Commons Attribution (CC BY) license (http://creativecommons.org/licenses/by/4.0/).

Article

Plasticized PVC Membrane Modified Electrodes: Voltammetry of Highly Hydrophobic Compounds

Ernő Lindner [1,*], Marcin Guzinski [2], Bradford Pendley [1] and Edward Chaum [2]

[1] Department of Biomedical Engineering, The University of Memphis, Memphis, TN 38152, USA; bpendley@memphis.edu
[2] Vanderbilt Eye Institute, Vanderbilt University Medical Center, Nashville, TN 37232, USA; marcin.guzinski@vumc.org (M.G.); edward.chaum@vumc.org (E.C.)
* Correspondence: elindner@memphis.edu

Received: 24 July 2020; Accepted: 21 August 2020; Published: 27 August 2020

Abstract: In the last 50 years, plasticized polyvinyl chloride (PVC) membranes have gained unique importance in chemical sensor development. Originally, these membranes separated two solutions in conventional ion-selective electrodes. Later, the same membranes were applied over a variety of supporting electrodes and used in both potentiometric and voltammetric measurements of ions and electrically charged molecules. The focus of this paper is to demonstrate the utility of the plasticized PVC membrane modified working electrode for the voltammetric measurement of highly lipophilic molecules. The plasticized PVC membrane prevents electrode fouling, extends the detection limit of the voltammetric methods to sub-micromolar concentrations, and minimizes interference by electrochemically active hydrophilic analytes.

Keywords: chemically modified electrodes; membrane-coated voltammetric sensors; antidepressant and immunosuppressant drugs; detection limit; resolution

---

1. Introduction

There are a few areas of analytical chemistry where conventional methods have been almost completely displaced by electrochemical sensor-based measurements [1,2]. In these areas, the electrochemical sensors were integrated into fully automated clinical laboratory analyzers, e.g., ion-selective electrodes (ISEs), and/or incorporated into hand-held instruments for short turnaround time (STAT) point of care testing (POCT) devices, e.g., sensors for measuring blood glucose concentrations [3]. The most essential parts of the sensors used in these applications are unique membranes that, beyond individual measurements in small sample volumes, also permit interference-free continuous monitoring of essential analytes in complex biological matrices like sweat, urine and whole blood [4,5]. However, the role and function of the membranes utilized in the various sensors are often fundamentally different. For example, in potentiometric ISEs, the membrane is the sensing element of the sensor that provides the analytically relevant, concentration-dependent signal [6]. In contrast, in voltammetric sensors, various membranes and coatings are used to enable or improve the analytical signal of a Pt, Au or glassy carbon (GC) working electrode. The latter is the concept of chemically modified electrodes (CMEs) [7,8]. By selecting the proper modifications, desirable properties (e.g., selectivity or detection limit) may be achieved, and electrode fouling can be prevented. The goal of this mini-review is to show the evolution of a new class of voltammetric sensors with unique selectivities and exceptional detection limits that are based upon sensor structures and membranes, which are most commonly utilized in solid contact ISEs. In line with this goal, we first describe the structure of membrane separated, conventional electrochemical cells with symmetric and asymmetric contacts that are used for potentiometric measurements of anions and cations (Schemes 1 and 2). Next, we show the similarities in the layer structure of these electrochemical cells and cells utilizing membrane-coated working electrodes for the measurement of redox-inactive cations and anions by ion transfer voltammetry (Scheme 3b). Finally, through selected examples, we demonstrate that the membrane coating on the working electrode surface can be utilized for the selective extraction and simultaneous measurement (oxidation or reduction) of highly hydrophobic, electrochemically active compounds (Scheme 3a).

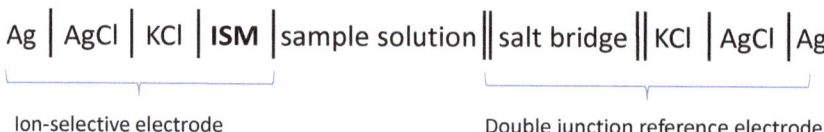

Scheme 1. Electrochemical notation of a potentiometric cell with a conventional, liquid inner contact ion-selective membrane electrode as indicator electrode and a double junction reference electrode.

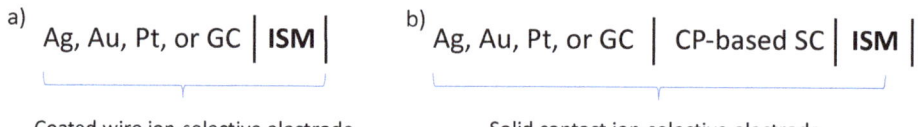

Scheme 2. Electrochemical notation of solid contact ion-selective electrodes. (**a**) Coated wire electrode; (**b**) Conductive polymer (CP)-based solid contact electrode.

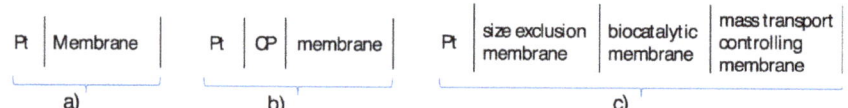

Scheme 3. Electrochemical notation of membrane modified voltammetric working electrodes. (**a**) Membrane directly deposited on the substrate electrode surface; (**b**) A conductive polymer (CP) layer is sandwiched between the substrate electrode and the membrane; (**c**) Multiple membrane layer modified electrode.

In conventional potentiometric sensors, the sensing membrane separates two solutions, the sample and the filling solution of the sensor, and the membrane potential is measured at zero current between two reference electrodes on the opposite sides of the membrane (Scheme 1). In 1971, Cattrall and Freiser [9] designed a simple compact electrode by dip-coating a Pt wire with ion-selective membrane (ISM) (Scheme 2a, coated wire electrode). Since the interface between the electron-conducting metal and the ion-conducting membrane in coated wire-type electrodes is not thermodynamically well defined (blocked interface) [10], these electrodes often have significant drift. The drift may also be related to the detachment of the membrane from the metal electrode surface and the formation of a thin water layer between the substrate electrode and the membrane [11,12]. To overcome some of the deficiencies of the coated wire electrode, an ion-to-electron transducer layer, generally a conductive polymer (CP) layer, has been sandwiched between the electron-conducting substrate and the ISM (Scheme 2b, solid contact ion-selective electrodes). In solid contact ISEs, the ISM is deposited directly over the conductive polymer by drop-casting or spin-coating (double polymer membrane modified electrodes). In electrochemical cells utilizing solid contact ISEs, the membrane potential is measured between the solid contact and the external reference electrode placed in the sample solution. The electrochemical notation of solid contact ISEs is shown in Scheme 2. In Schemes 1 and 2, ISM represents the ion-selective membrane, CP stands for conductive polymer, vertical lines represent phase boundaries, and two vertical lines represent a liquid/liquid junction. Under ideal conditions, each phase boundary is in thermodynamic equilibrium and the composition of the contacting phases, including the ion-selective membrane (ISM), is constant [6].

The layer structure of membrane-modified voltammetric sensors can be the same as that of solid contact ion-selective electrodes (Scheme 3a,b) or significantly more complex (Scheme 3c). The more complex layer structure is needed to enable the voltammetric sensors to be used for quantitative measurements in complex matrices. Although the oxidation and reduction of organic molecules on electrode surfaces offer almost limitless possibilities for both batch [13] and flow-analytical [14,15] measurements for a variety of analytes, beyond some exceptions, voltammetric sensors are mainly used in research laboratories [16]. The limited spread of practical analytical methods based on voltammetric sensors can be traced back to three main causes: (i) limited selectivity of voltammetric measurements and the difficulty of eliminating the influence of interfering compounds on the voltammetric signal; (ii) inadequate detection limit of the voltammetric methods and the challenges of quantification of analytes at sub-micromolar concentration; (iii) inconsistency of the voltammetric signals due to electrode fouling. Since the voltammetric signal (the measured current) is a function of the working electrode surface area, contaminations of the electrode surface from the sample or as a consequence of the electrochemical reactions may contribute to a negative or positive bias in the analytical signal. When resistive layers build up on or around the sensing surface, the sensor may lose its utility because its signal deteriorates and/or its response slows down. The buildup of resistive layers and encapsulation of sensors is most relevant during in-vivo applications when the sensors are implanted into tissues for monitoring. The issue is related to the biocompatibility of the sensors and the "sensor compatibility" of the environment where it is working [16–18]. While unwanted precipitation or adsorption processes may reduce the response of a voltammetric working electrode, targeted manipulation of the surface chemistries can enhance its electrochemical response, e.g., improve the selectivity or detection limit, prevent electrode fouling, etc. However, the "proper modifications" may require multi-layer constructions, as shown in Scheme 3b,c. For example, in ion-transfer voltammetry, double polymer membrane modified electrodes are used, [19–21] and the Pt working electrode surface of the voltammetric enzyme sensor for the measurement of glucose in whole blood is generally coated by a selectivity enhancing layer, a biocatalytic layer and a mass transport controlling layer. Each of these layers have specific roles for adequate voltammetric response.

Within the large family of CMEs, voltammetric working electrodes with an organic film coating on their surfaces represent a special group. When the aqueous phase (sample) contains hydrophilic electrolyte and the organic film is loaded with a hydrophobic electrolyte, then the organic film/sample solution interface behaves analogously to polarizable working electrodes which can be used to drive ions across the aqueous/organic interface. The analytical signal is the current related to the ion-transfer of the measured ions across the aqueous/organic interface. The analytical methods utilized in combination with double membrane-coated working electrodes (Scheme 3b) evolved from ion-transfer voltammetry at the interface between two immiscible electrolyte solutions (ITIES) [20]. ITIES was developed for the measurement of redox-inactive, generally hydrophilic anions and cations [22–24], but has also been utilized for the voltammetric ion transfer of biologically relevant polyions such as heparin [25,26] and protamine [27]. ITIES is attractive because it enables the measurements of ions without their electrolysis [20,28]. With the double membrane arrangement, i.e., a conductive polymer-supported thin plasticized PVC membrane, as shown in Scheme 2b, the oxidation or reduction of the conductive polymer drives the transfer of anions or cations into the PVC membrane, respectively (Figure 1). The selectivity of the method is related to differences in the solvation energies of the different ions which can be facilitated by incorporating selective ionophores and or ion-exchange sites in the polymeric membranes [22–24] and enhanced by kinetic effects of ion hydrophobicity [29]. Using an adequate applied potential, the analyte ions can be concentrated in the organic phase. Following concentration, the ions can be stripped from the organic phase into the aqueous phase using a voltammetric technique, e.g., linear sweep voltammetry. By using a highly plasticized, ionophore loaded ion-selective membrane as the organic phase over a conductive polymer layer for ITIES, as in Scheme 2 and Figure 1, in combination with stripping voltammetry, subnanomolar concentrations can be measured for both cationic [30] and anionic [21,23] analytes.

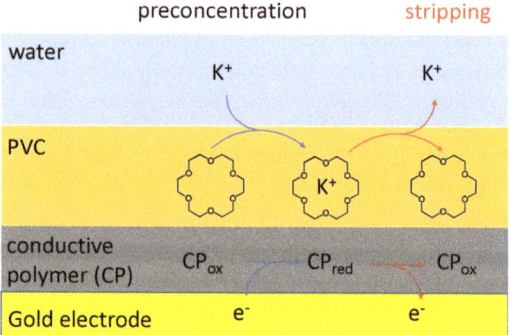

Figure 1. Schematic representation of the processes during the concentration (black arrows) and stripping (red arrows) of cations using a conductive polymer/plasticized PVC membrane modified electrode. The 18-crown-6 molecule in the PVC membrane represents any selective ionophore which may be used in this arrangement.

In contrast to the ion-transfer voltammetric methods, the single membrane-coated working electrode (Scheme 3a) is used for the measurement of electrochemically active highly lipophilic analytes [31,32]. The highly lipophilic analytes spontaneously partition into the membrane where their oxidation/reduction provides the analytical signal. Consequently, these measurements can be considered as voltammetric determinations in a non-aqueous phase, i.e., within the membrane coating. Accordingly, potential sweep and potential step methods, with or without the integration of the recorded current, can all be used in combination with this electrode. The selectivity and detection limit of this working electrode is related to the membrane/solution partition coefficient of the analyte and interfering compounds.

2. Materials and Methods

2.1. Membranes and Membrane Deposition

For the preparation of membranes poly(vinyl chloride) (PVC, high molecular weight), and bis(2-ethylhexyl)sebacate (DOS) were purchased from Sigma-Aldrich, St. Louis, MO, USA. Tetradodecylammonium tetrakis(pentafuorophenyl)borate (TDDA-TPFPhB) was prepared by metathesis reaction between tetradodecylammonium chloride (TDDACl) (Sigma-Aldrich, St. Louis, MO, USA) and high purity sodium tetrakis-(pentafluorophenyl)borate (KTPFPhB) (Boulder Scientific Company, Mead, CO, USA) in dichloromethane. PVC membrane solutions were prepared by dissolving 25 mg PVC, 50 mg DOS, 22 mg TDDA-TPFPhB and 3 mg NaTPFPhB in 1 mL freshly distilled of tetrahydrofuran (THF) (Sigma-Aldrich, St. Louis, MO, USA). This membrane cocktail was used to spin-coat the surface of a 3 mm diameter glassy carbon (GC) electrode (Model MF-2012, BASi, West Lafayette, IN, USA). Before the membrane deposition, the surface of the GC electrode was polished on wet microcloth pads using Al_2O_3-based slurry (Mager Scientific, Dexter, MI, USA) with grain sizes of 1, 0.3 and 0.05 μm [31]. For spin-coating, the electrode was secured in the three-jaw chuck of a Dewalt 20V MAX drill press (Baltimore, MD, USA) and dipped into a PVC membrane cocktail. After the removal of the electrode from the cocktail, it was rotated for 60 s at 2000 rpm and left in an upright position until the complete evaporation of THF. This protocol resulted in 1–3 μm thick PVC membrane-coating on the electrode surface. For spin-coating the CHI Model 131 electrode, which was used in the CHI Model 130 flow cell, a KW-4 spin coater (SETCAS LLC, San Diego, CA, USA) was employed at 1650 RPM rotation rate. While the GC electrode was rotated, 50 μL of the membrane solution was dropped on the center of the electrode. The GC electrode was rotated for 3 min and then was left for 3 h for complete THF evaporation.

2.2. Batch Measurements with Linear Sweep Voltammetry and Chronoamperometry

Linear sweep voltammetry (LSV), and chronoamperometry (CA) experiments using the PVC membrane-coated GC electrodes were performed in pH ~7.2 phosphate buffered saline (PBS) buffer solution with 0.1 mol/L potassium chloride (KCl) (Sigma-Aldrich, St. Louis, MO, USA) or whole blood, with ~0.17% Ethylenediaminetetraacetic acid (EDTA) as anticoagulant, using a GC rod and a single junction silver/silver chloride (Ag/AgCl) 1 mol/L KCl reference electrode as the counter and reference electrodes, respectively. The PBS buffer (pH ~7.2) was prepared as a mixture of 0.1 mol/L KH_2PO_4 (Sigma-Aldrich, St. Louis, MO), 0.1 mol/L K_2HPO_4 (Sigma-Aldrich, St. Louis, MO), 0.1 M KCl, and 0.045 mol/L NaOH (Sigma-Aldrich, St. Louis, MO). The PBS buffer solutions were spiked with stock solutions of ascorbic acid (AA)(Acros Organics, a Thermo Fisher Scientific Subsidiary, Waltham, MA, USA), p-acetamino phenol (APAP) (Sigma-Aldrich, St. Louis, MO, USA) and the individual drugs. The LSV scans were performed between 0 and 1.4 V at 0.1 V/s scan rate. The CA experiments were performed at the peak potentials determined in the LSV experiments for the individual drugs. The calibration curves were recorded by spiking a PBS background electrolyte or whole blood by stock solutions. For the voltammetric measurements, a CH Instruments (Austin, TX, USA) model CHI 842B potentiostat was used with CH Instruments software version 18.01.

2.3. Chronoamperometry in a Continuous Flow Analysis Mode

For the continuous flow analysis (CFA) measurements, a CH Instruments model CH130 thin layer flow cell was used. In the thin layer cell, the PVC membrane-coated GC working electrode was used in combination with an Ag|AgCl|3.0 mol/L sodium chloride (NaCl) reference electrode and a stainless steel counter electrode. The cell was connected with a Gilson Minipuls 3 peristaltic pump and Hamilton modular valve positioner. The standard solutions in 0.1 mol/L, pH ~7.2 PBS were transported by 0.5 mL/min flow rate.

3. Results

Voltammetric Measurement of Highly Lipophilic Molecules

In voltammetric analysis, the modification of the working electrode surface may be motivated by various reasons, e.g., to improve the selectivity, sensitivity and detection limit of the method [33,34]. By coating a working electrode surface with an organic film, e.g., a plasticized PVC membrane, analyte molecules spontaneously partition into the membrane. For highly lipophilic analytes, it means that the analyte concentration in the membrane can be orders of magnitudes larger than in the aqueous sample. Since the analytically relevant signal of the membrane-coated sensor is proportional to the analyte concentration in the membrane, the attainable detection limit with the membrane-coated sensor is expected to be much lower compared to measurements with an unmodified working electrode. The extension of the detection limit towards lower concentrations can be estimated from the current ratio (or slope/sensitivity ratio), which is expected at a given concentration with or without coating the working electrode surface. The correlation between the measured current and the concentration in linear sweep voltammetry (LSV) using a macroelectrode is described by Equation (1) (Randles–Sevcik equation) and for micro disc electrodes at steady state by Equation (2). The corresponding current ratios are provided by Equations (3) and (4), respectively.

$$i_{p,w} = (2.69 \times 10^5) n^{3/2} A D_w^{1/2} c_w v^{1/2} \tag{1}$$

$$i_{L,w} = 4nFD_w r c_w \tag{2}$$

$$\frac{i_{p,m}}{i_{p,w}} = \frac{D_m^{1/2} c_m}{D_w^{1/2} c_w} \tag{3}$$

$$\frac{i_{L,m}}{i_{L,w}} = \frac{D_m c_m}{D_w c_w} \tag{4}$$

where $i_{p,w}$, $i_{p,m}$ and $i_{L,w}$, $i_{L,m}$ are the peak or limiting currents in amperes, respectively, n number of electrons involved in the electrochemical reaction, F (C/mol) is the Faraday number, A (cm^2) is the surface area of the electrode, r (cm) is the radius of the micro disc electrode, D (cm^2/s) is the diffusion coefficient, c (mol/cm^3) is the concentration and v (V/s) is the scan rate. The subscripts w or m indicate the currents, concentrations, and diffusion coefficients in water or in the membrane phase.

The analyte concentration in the membrane can be estimated using the octanol/water partition coefficient ($P_{o/w}$):

$$P_{o/w} = \frac{c_o}{c_w} \tag{5}$$

where c_o and c_w are the concentrations of an analyte in n-octanol and water. As a consequence of the difference in the current versus concentration relationship for macro and microelectrodes (Equations (1) and (2)), the combination of Equation (5) with Equations (3) and (4) shows a significant difference in the expected signal amplification with a membrane-coated macro (Equation (6)) or micro (Equation (7)) electrode. Related to Equations (6) and (7), it must be emphasized that for more realistic estimates of the signal amplification, $P_{o/w}$ should be replaced by $P_{m/w}$, the membrane-water partition coefficient.

$$\frac{i_{p,m}}{i_{p,w}} = \frac{D_m^{1/2} c_m}{D_w^{1/2} c_w} = P_{o/w} \sqrt{\frac{D_m}{D_w}} \tag{6}$$

$$\frac{i_{L,m}}{i_{L,w}} = \frac{D_m c_m}{D_w c_w} = P_{o/w} \frac{D_m}{D_w} \tag{7}$$

As an example, Equations (6) and (7) were applied to calculate the expected signal amplification for a lipophilic and a hydrophilic drug. The highly lipophilic anesthetic drug propofol ($logP_{o/w}(Propofol) \approx 3.8$) and p-acetamino phenol (APAP) ($logP_{APAP} \approx 0.31$), a potential interfering compound during propofol measurement, were selected as characteristic examples. As a diffusion coefficient in the aqueous solution, $D_w \approx 7 \times 10^{-6}$ cm^2/s was used [35]. For the diffusion coefficients in the membrane, $D_m = 2 \times 10^{-8}$ cm^2/s and $D_m = 2 \times 10^{-7}$ cm^2/s were used [36,37]. These values were reported as ionophore diffusion coefficients in the ion-selective membranes [36] or calculated from the scan rate dependence of the peak current (Equation (1)) [37] in plasticized PVC membranes with 1 to 2 PVC to plasticizer ratio. The results of these calculations are summarized in Table 1.

Table 1. Peak current ratios ($i_{p,m}/i_{p,w}$) which would be expected if linear sweep voltammetry (LSV) experiments were performed with or without coating the working electrode surface with a plasticized PVC membrane in solutions of the same concentration.

	($i_{p,m}/i_{p,w}$)$_{propofol}$		($i_{p,m}/i_{p,w}$)$_{APAP}$	
	$D_m=2\times10^{-8}$ cm^2/s	$D_m=2\times10^{-7}$ cm^2/s	$D_m=2\times10^{-8}$ cm^2/s	$D_m=2\times10^{-7}$ cm^2/s
Macroelectrode, Equation (6)	337	1066	0.11	0.35
Microelectrode, Equation (7)	18.0	180	0.006	0.06

As shown in the Table 1, a remarkable signal amplification can be achieved ($i_{p,m}/i_{p,w} \gg 1$) with highly lipophilic drugs (analyte with large $P_{o/w}$), while the signal for hydrophilic drugs is reduced compared to an uncoated electrode. The amplification of the analyte signal increases the signal to noise and signal to background current ratios. Thus, it can be assumed that the signal amplification is inversely proportional to the detection limit. Consequently, detection limits in the lower nanomolar concentrations may be attainable with voltammetric methods utilizing organic membrane-coated working electrodes. However, there are significant differences in the expected signal amplification with macro and microelectrodes. While microelectrodes have unique advantages for voltammetric measurements in highly resistive media, like the plasticized polymeric membranes (small IR potential drop), the detection limit using LSV is expected to be better with the membrane-coated macroelectrode. The difference is related to the influence of the diffusion coefficient on the current signal, i.e., $\sqrt{D_m}$ for macroelectrodes but D_m for microelectrodes.

On the other hand, the discrimination against hydrophilic interferences is expected to be the same with the membrane-coated macro and microelectrode. It depends only on the partition coefficients of the drugs. Using the octanol/water partition coefficient of propofol and APAP, we get:

$$discrimination = \frac{P_{o/w}^{propofol}}{P_{o/w}^{APAP}} = \frac{6309}{2.04} = 3090$$

The discrimination can also be determined experimentally from the $i_{p,m}/i_{p,w}$ or calibration slope ratios for the analyte and the interfering compound. Using the data in columns 2 and 4 of Table 1 provides the same results for both the macro and the microelectrode:

$$discrimination = \frac{\left(\frac{i_{p,m}}{i_{p,w}}\right)_{propofol}}{\left(\frac{i_{p,m}}{i_{p,w}}\right)_{APAP}} = \left(\frac{337.26}{0.109}\right)_{macroelectr.} = \left(\frac{18.03}{0.0058}\right)_{microelectr.} = 3090$$

In Figure 2, linear sweep voltammograms recorded with a bare and a membrane-coated electrode in ascorbic acid (AA) and p-acetamino phenol (APAP) solutions are shown to demonstrate the effectiveness of the organic membrane coating for minimizing the interference of electrochemically active, hydrophilic compounds on the voltammetric signal. As seen in the insets, 49.8 µmol/L AA did not result in any measurable current signal with the PVC membrane-coated electrode, while in the 117.6 µmol/L solution of APAP solutions, the current signal was about 20 times lower than with the bare electrode. The difference between AA and APAP is related to the difference in their partition coefficients [38]: logP$_{AA}$ = −1.84 and logP$_{APAP}$ = 0.31.

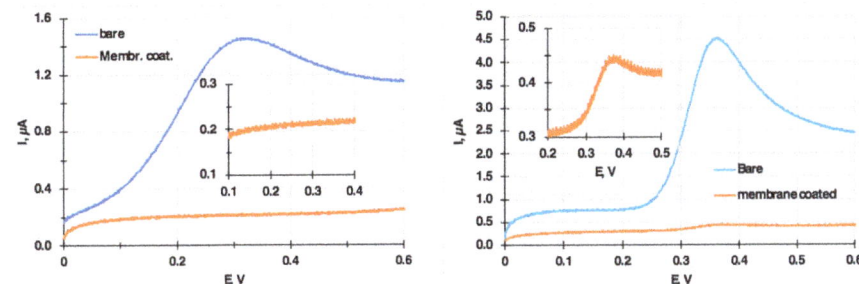

Figure 2. Linear sweep voltammograms recorded with a bare and a membrane-coated electrode in 49.8 µmol/L ascorbic acid (AA) (**left**) and 117.6 µmol/L p-acetamino phenol (APAP) (**right**) solutions. The insets show the membrane-coated sensor signal with much higher resolution in the current scale. The PVC membrane composition is 25% PVC, 50% bis(2-ethylhexyl)sebacate (DOS), 22% tetradodecylammonium tetrakis(pentafuorophenyl)borate (TDDA-TPFPhB) and 3% sodium tetrakis[3,5bis(trifluoromethyl) phenyl] borate (NaTFPhB).

In Figure 3, the processes during voltammetric measurements of propofol with the PVC membrane-modified electrode are summarized. The plasticized PVC membrane used for coating the GC electrode surface was similar to membranes used in ionophore-based ion-selective electrodes and for ITIES measurement of charged molecules [19,21,23]. However, when these membranes are used in voltammetric measurements, they must also contain a lipophilic background electrolyte, e.g., tetradodecylammonium tetrakis(pentafluorophenyl) borate (TDDATPFPhB). These membranes are labeled as "liquid membranes" due to their very high plasticizer content (up to 86%) [39] in which the diffusion coefficients are similar to viscous solutions. Highly lipophilic compounds like propofol preferentially partition in the membrane plasticizer in which the voltammetric measurement is performed. The enrichment of the analyte in the plasticized PVC membrane is controlled by the partition coefficient of the analyte. The logarithm of the octanol/water partition coefficient ($logP_{o/w}$) of propofol is between 3.83 and 4.15, i.e., the concentration of propofol is about 10,000 times higher in the membrane where the electrochemical reaction occurs than in the aqueous sample solution. At the same time, the hydrophilic interferences, such as ascorbic acid ($logP_{AA}$ = −1.84) [38] and p-acetamino phenol ($logP_{APAP}$ = 0.31) [38], are minimally extracted into the membrane phase, i.e., their concentration generally is negligible in the membrane, which explains the LSVs in Figure 2.

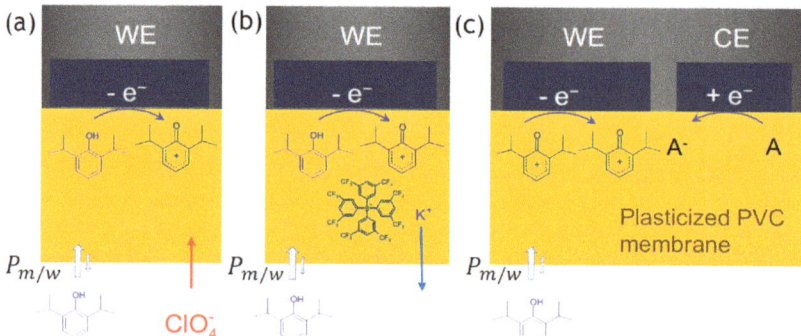

Figure 3. Three different variations of the processes during voltammetric measurements of propofol with the PVC membrane modified electrode. In all three examples, propofol preferentially partitions into the membrane, and during its electrochemical oxidation, propofol positively charged intermediates are generated. The electroneutrality in the membrane is sustained through (**a**) the uptake of hydrophobic counterions, e.g., (ClO_4^-) from the solution; (**b**) the release of positively charged counterions of the cation exchanger (e.g., K^+) incorporated in the membrane; (**c**) the generation of negatively charged species on the counter electrode. In this scenario, both the working (WE) and counter (CE) electrodes are coated with the membrane.

Figure 4 shows the response of the membrane-coated propofol sensor during calibrations by chronoamperometry and linear sweep voltammetry. The attainable detection limit of the membrane-coated propofol sensor was calculated from the standard deviation of the noise (SD_{bkg}) in the background electrolyte during linear sweep voltammetric and chronoamperometric measurements and the slopes of the corresponding calibration curves (S) using the formula:

$$LOD = 3 \times SD_{bkg}/S$$

In addition, we assessed the resolution of the propofol measurement using the residual mean standard deviation (RMSD) of the data points around the line fitted to the data points of the calibration curve, instead of the standard deviation of the background noise. The resolution is defined as a minimum concentration difference between two solutions essential for their quantitative determination.

$$Resolution = 3 \times RMSD/S$$

The calculated values in the presence of different background electrolytes and using calibration data points recorded in narrower or broader concentration ranges are summarized in Table 2 [31,40,41].

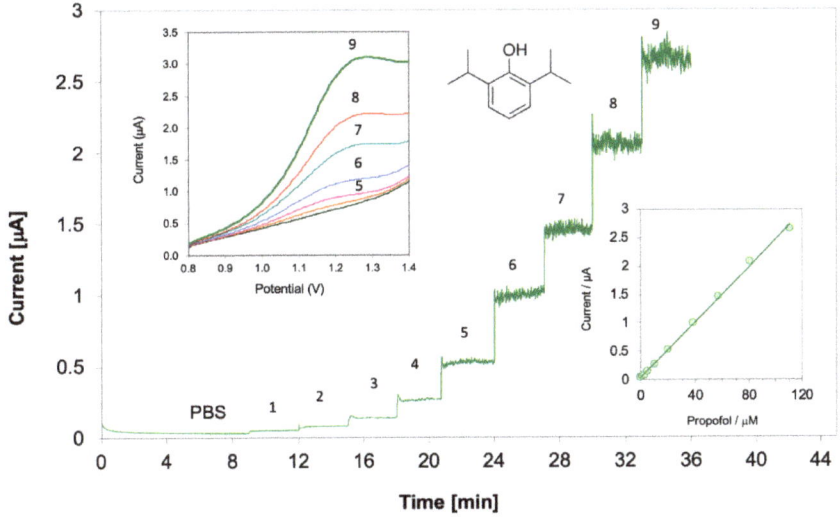

Figure 4. Chronoamperometric response of a PVC-membrane-coated glassy carbon (GC) electrode upon spiking the phosphate buffered saline (PBS) background electrolyte with propofol standard solutions. The concentrations in the PBS solution labeled 1 to 9 are 1.25, 2.5, 4.98, 9.9, 19.6, 38.4, 56.6, 80.5 and 111.1 µmol/L. The inset in the lower right is the calibration curve constructed from the chronoamperometric currents. The inset in the upper left corner shows linear sweep voltammograms recorded in PBS background solution and in PBS with propofol concentrations between 9.9 and 111.1 µmol/L. Membrane composition 25.5% PVC, 50.9% 2-nitrophenyl octyl ether (o-NPOE), 21.2% tetradodecylammonium tetrakis(pentafluorophenyl) borate (TDDATPFPhB), 2.4% sodium tetrakis[3,5bis(trifluoromethyl) phenyl] borate (NaTFPhB).

Table 2. Detection limit (LOD) and resolution values calculated for the propofol sensor from triplicate chronoamperometry (CA) studies in different background electrolytes. The background electrolytes are prepared in PBS containing 3 mmol/L ascorbic acid (AA), 1 mmol/L p-acetamino phenol (APAP), 5% w/v bovine serum albumin (BSA) or a combination of all three of these potential interferents (mix).

Membrane Composition *	Background	Concentration Range † [µmol/L]	LOD [µmol/L]	Resolution [µmol/L]
o-NPOE plasticizer	PBS	0–56.6	0.03 ± 0.01	1.1 ± 0.2
	3 mmol/L AA	0–56.6	0.05 ± 0.05	2.0 ± 1.0
	1 mmol/L APAP	0–56.6	0.08 ± 0.02	4.6 ± 0.9
	5% BSA	5–56.6	2.2 ± 3.1	14 ± 2
	mix	2.5–109.8	0.5 ± 0.4	28 ± 5
DOS plasticizer	PBS	0–56.6	0.12 ± 0.05	4.3 ± 0.4
	mix	0–56.6	3.0 ± 0.3	4.5 ± 2.3

* The composition of the PVC membranes are 25.5% PVC, 50.9% 2-nitrophenyl octyl ether (o-NPOE), 21.2% tetradodecylammonium tetrakis(pentafluorophenyl) borate (TDDATPFPhB), 2.4% sodium tetrakis[3,5bis-(trifluoromethyl) phenyl] borate dihydrate (NaTFPhB) and 25.1% PVC, 49.9 bis(2-ethylhexyl)sebacate (DOS), 22.6% TDDATPFPhB and 2.4% potassium tetrakis-(pentafluorophenyl) borate (KTPFPhB). † The slope and RMSD value of the calibration curve was calculated from data points recorded in the concentration range.

As shown in Table 2, by coating the GC working electrode surface with the plasticized PVC film, the detection limit in propofol measurements can be extended below the lower limit of the therapeutic concentration range (1 and 60 µmol/dm^3). The interference induced by electrochemically oxidizable species are negligible (AA) or small (APAP), even if those are at the maximum of their therapeutic concentrations. From the data, it appears that by optimizing the membrane composition, i.e., the selection of the membrane plasticizer, lipophilic background electrolyte and ion-exchanger, and their concentration in the membrane, both the selectivity and the detection limit of the sensor may be improved. Finally, it is important to note that in contrast to aqueous electrolyte solutions, during the chronoamperometric determination of propofol with the PVC membrane-coated sensor, no electrode fouling is observed [31].

Based on Equations (6) and (7), the signal amplification of the organic film (PVC membrane-coated) voltammetric sensor is directly proportional with the partition coefficient of the target analyte between the membrane and aqueous phase, i.e., the lowest detection limits are expected during the voltammetric determination of the most lipophilic compounds. On the other hand, the selectivity of the membrane modified electrode is controlled by the partition coefficient ratio of the analyte and the interfering compound. Consequently, the voltammetric determinations with the membrane modified electrodes are most desirable for tasks requiring the measurement of sub-micromolar concentrations of lipophilic analytes in the presence of large excess hydrophilic interferents. A list of electrochemically active target analytes (antidepressant and immunosuppressant drugs) are summarized in Table 3. All these drugs are highly lipophilic with therapeutic blood concentration ranges below 1.0 µmol/L, which could be measured in the presence of up to 1.3 mmol/L APAP and 0.06 mmol/L AA.

Table 3. List of highly lipophilic, electrochemically active analytes (antidepressant drugs) with sub-micromolar therapeutic concentration ranges as potential targets for the development of voltammetric methods based on organic film-coated working electrodes. The log p values are generally from DRUGBANK (www.drugbank.ca). Experimentally determined values are provided with the source. SSRI stands for selective serotonin reuptake inhibitors.

Sertraline (Zoloft) Antidepressant SSRI		log P = 5.06–5.15 MW: 306.229 g/mol Concentration range: 30–200 ng/mL 0.1–0.65 µmol/L
Amitriptyline (Elavil) Antidepressant		log P = 4.81–5.1 log P_{exp} = 4.92 [42] MW: 277.403 g/mol Concentration range: 75–175 µg/L 0.27–0.63 µmol/L
Aripiprazole (Abilify) Antipsychotic		log P = 4.9–5.21 MW: 448.385 g/mol Concentration range: 150 and 300 ng/mL 0.33–0.67 µmol/L

Table 3. *Cont.*

Drug	Structure	Properties
Sirolimus (Rapamycin) Immunosuppressant	[structure]	log P = 4.8–7.5 MW: 914.2 g/mol Concentration range: 4–20 ng/mL 4.4–21.9 nmol/L
Everolimus (Zortress) Immunosuppressant	[structure]	log P = 5.0–7.4 MW: 958.240 g/mol Concentration range: 3–8 ng/mL 3.1–8.3 nmol/L
Citalopram (Celexa) Antidepressant SSRI	[structure]	log P = 3.6–3.8 MW: 324.392 g/mol Concentration range: 50–100 ng/mL 0.15–0.3 µmol/L

LSVs recorded with membrane-coated glassy carbon (GC) electrodes in sertraline, amitriptyline, aripiprazole, sirolimus, everolimus and citalopram solutions of different concentrations are shown in Figure 5. The peak currents of the LSVs were used to construct calibration curves for the individual drugs. Calibration curves were also recorded using a chronoamperometric measurement protocol by applying the drug-specific potential values to the PVC membrane-coated electrode. The chronoamperometric transients recorded during the calibration of the PVC membrane-coated electrode in sertraline solutions and the corresponding calibration curve is shown in Figure 6 as an example. Based on the calibration curves recorded with linear sweep voltammetry and chronoamperometry, we calculated the LOD and resolution of the determinations for all the drugs. The results of the calculations are summarized in Table 4. As shown in Table 4, both types of measurement methods result in sub-micromolar detection limits, and in the case of amitriptyline, even in whole blood. In agreement with the expectations, both the LOD and resolution values are better in chronoamperometric experiments. The detection limit and the resolution of the chronoamperometric method could be further improved by performing the measurements in a flow through manifold (not shown). The reproducibility of the chronoamperometric measurements in a continuous flow mode of measurement is shown in Figure 7.

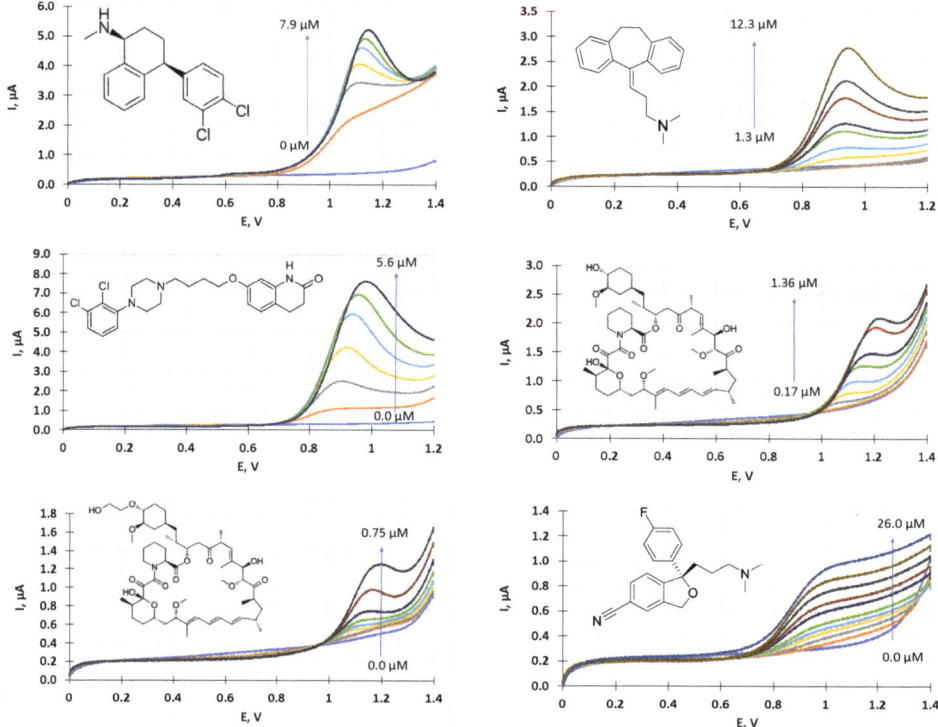

Figure 5. Linear sweep voltammograms recorded in sertraline (**upper left**), amitriptyline (**upper right**) and aripiprazone (**middle left**), sirolimus (**middle right**), everolimus (**lower left**) and citalopram (**lower right**) solutions. The concentration ranges are indicated in the individual pictures.

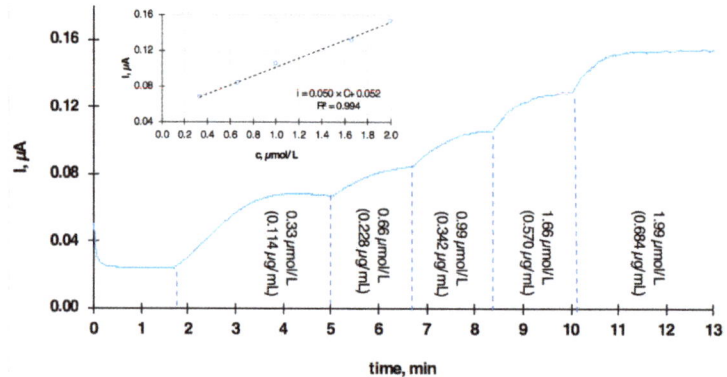

Figure 6. Chronoamperometric transients recorded with the plasticized PVC membrane-coated GC electrode in PBS solution of different sertraline concentrations. Applied potential: 1.0 V.

Table 4. Detection limit (LOD), resolution of the plasticized PVC membrane-coated GC electrode for amitriptyline, sertraline, aripiprazole, sirolimus, everolimus and citalopram.

Analyte	Method	Matrix	LOD µmol/L	Resolution µmol/L
Amitriptyline	LSV	PBS	0.09 ± 0.05 (n = 3)	0.60 ± 0.2 (n = 3)
		Whole blood	0.07 ± 0.03 (n = 2)	0.09 ± 0.03 (n = 2)
	CA	PBS	0.01 ± 0.03 (n = 5)	0.05 ± 0.02 (n = 3)
		Whole blood	0.03	0.07
Sertraline	LSV	PBS	0.13	0.75
	CA	PBS	0.003	0.12
Aripiprazone	LSV	PBS	0.02	0.43
	CA	PBS	0.009	0.07
Sirolimus	LSV	PBS	0.02	0.07
	CA	PBS	0.003	0.02
Everolimus	LSV	PBS	0.01	0.05
	CA	PBS	0.02	0.04
Citalopram	LSV	PBS	0.6	0.7
	CA	PBS	0.008	0.08

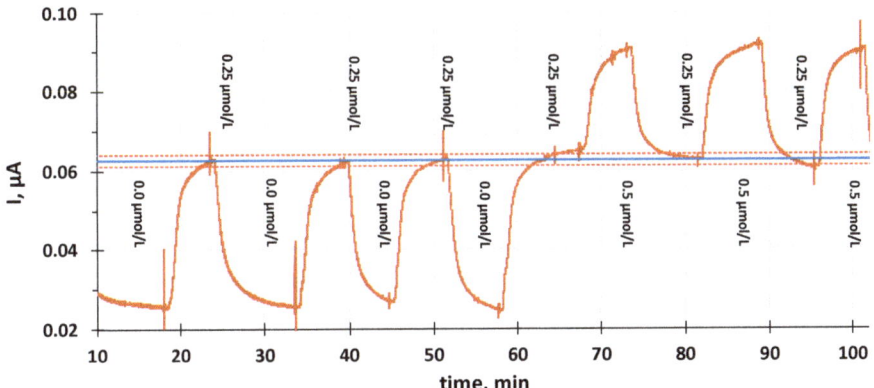

Figure 7. The reproducibility of the chronoamperometric measurements in a continuous flow mode of measurement of amitriptyline with the PVC membrane-coated GC electrode in PBS solution. Applied potential: 0.95 V. The standard solutions are switched between 0 and 0.25 µmol/L and 0.25 and 0.5 µmol/L in the PBS background. The blue line represents the mean value of the current in 0.25 µmol/L amitriptyline and the red dotted lines represent ±1 standard deviation range of the data around the mean.

Using a plasticized PVC membrane-coated glassy carbon (GC) electrode for the measurement of lipophilic drugs also prevented electrode fouling, which is commonly experienced with the uncoated electrode (Figure 8).

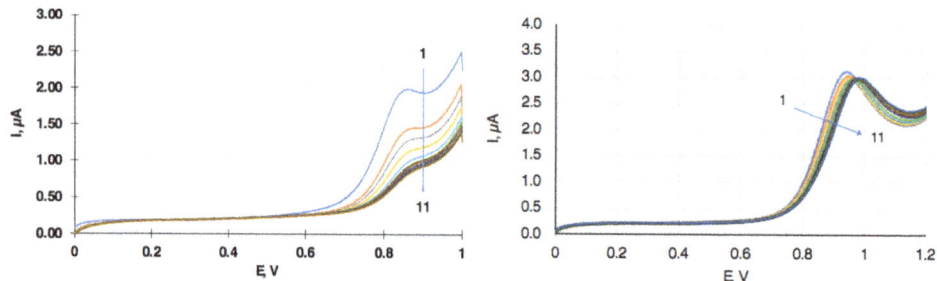

Figure 8. Repeated LSVs of a bare (**left**) and PVC membrane-coated (**right**) GC electrode in 50 μmol/L (bare, **left**) and 10 μmol/L (membrane-coated, **right**) amitriptyline containing solutions.

4. Conclusions

The simple modification of voltammetric working electrodes by coating their surfaces with a thin film of highly plasticized PVC membrane (an organic film) opens up the possibility of monitoring highly lipophilic drugs continuously in biological samples. In general, PVC membrane coating prevents electrode fouling and permits the quantification of analytes in sub-micromolar concentrations even in the presence of a large excess of electrochemically active, hydrophilic interfering compounds. The improvement in the sensitivity of the voltammetric measurements with the PVC membrane-coated working electrode, compared to measurements with conventional working electrodes without coating, is a function of the increased concentration and decreased diffusion coefficient ($\sqrt{D}$ or D) of the analyte in the membrane in which the voltammetric measurement is performed. For the lipophilic analytes discussed in this paper, the concentration increase is estimated between 10^4 and 10^7 (from $logP_{o/w} = 4$ to $logP_{o/w} = 7$). At the same time, the diffusion coefficients in the highly plasticized PVC membranes are about 2–2.5 orders of magnitude smaller than in aqueous solutions. As a consequence, a three orders of magnitude signal amplification and a three orders of magnitude improvement in the detection limit can be realized. The selectivity improvement with the membrane-coated voltammetric sensor is a function of the partition coefficients for the analyte and the interfering compound. Consequently, hydrophilic analytes at their therapeutic concentration, e.g., ascorbic acid and p-acetamino phenol, have no or only negligible effects on the plasticized PVC membrane-coated sensor signal.

Author Contributions: Conceptualization, E.L. & E.C.; methodology, E.L. & M.G.; investigation, M.G.; writing—review and editing, E.L. & B.P.; funding acquisition, E.C. All authors have read and agreed to the published version of the manuscript.

Funding: This research was funded by the Telemedicine and Advanced Technology Research Center (TATRC) at the US Army Medical Research and Materiel Command under Awards W81XWH-05-2-0064, W81XWH-10-1-0358 and W81XWH-13-C-0099. The views, opinions and/or findings contained in this report are those of the author(s) and should not be construed as an official Department of the Army position, policy, or decision unless so designated by other documentation. These studies were also supported in part by research grants from the Tennessee Technology Development Corporation, a FedEx Institute of Technology SENSORIUM Grant, unrestricted departmental grants from Research to Prevent Blindness (New York, NY), NIH EY024063, and by the Margy Ann and J. Donald M. Gass Chair at Vanderbilt University Medical Center.

Conflicts of Interest: Lindner and Chaum are authors on issued patents related to this technology. Chaum has a declared equity interest in Infusense, a startup company with the license to commercialize the technology presented in this manuscript.

References

1. Zdrachek, E.; Bakker, E. Potentiometric Sensing. *Anal. Chem.* **2019**, *91*, 2–26. [CrossRef] [PubMed]
2. Lindner, E.; Buck, R.P. Microfabricated Potentiometric Electrodes and their In Vivo Application. *Anal. Chem.* **2000**, *72*, 336A–345A. [CrossRef] [PubMed]
3. Abbot Point of Care The i-STAT System. Available online: https://www.pointofcare.abbott/us/en/home (accessed on 7 March 2020).

4. Phillips, F.; Kaczor, K.; Gandhi, N.; Pendley, B.D.; Danish, R.K.; Neuman, M.R.; Toth, B.; Horvath, V.; Lindner, E. Measurement of Sodium Ion Activity in Undiluted Urine with Cation-Selective Polymeric Membrane Electrodes after the Removal of Interfering Compounds. *Talanta* **2007**, *74*, 255–264. [CrossRef]
5. Sempionatto, J.R.; Jeerapan, I.; Krishnan, S.; Wang, J. Wearable Chemical Sensors: Emerging Systems for On-Body Analytical Chemistry. *Anal. Chem.* **2020**, *92*, 378–396. [CrossRef] [PubMed]
6. Lindner, E.; Pendley, B.D. A tutorial on the application of ion-selective electrode potentiometry: An analytical method with unique qualities, unexplored opportunities and potential pitfalls; Tutorial. *Anal. Chim. Acta* **2013**, *762*, 1–13. [CrossRef]
7. Murray, R.W.; Ewing, A.G.; Durst, R.A. Chemically Modified Electrodes—Molecular Design for Electroanalysis. *Anal. Chem.* **1987**, *59*, A379. [CrossRef]
8. Guadalupe, A.R.; Abruña, H.D. Electroanalysis with Chemically Modified Electrodes. *Anal. Chem.* **1985**, *57*, 142–149. [CrossRef]
9. Cattrall, R.W.; Freiser, H. Coated wire ion selective electrodes. *Anal. Chem.* **1971**, *43*, 1905–1906. [CrossRef]
10. Buck, R.P.; Lindner, E. Studies of potential generation accross membrane sensors at interfaces and through bulk. *Acc. Chem. Res.* **1998**, *31*, 257–266. [CrossRef]
11. Fibbioli, M.; Morf, W.E.; Badertscher, M.; de Rooij, N.F.; Pretsch, E. Potential drifts of solid-contacted ion-selective electrodes due to zero-current ion fluxes through the sensor membrane. *Electroanalysis* **2000**, *12*, 1286–1292. [CrossRef]
12. Hambly, B.; Guzinski, M.; Pendley, B.; Lindner, E. Evaluation, pitfalls and recommendations for the "water layer test" for solid contact ion-selective electrodes. *Electroanalysis* **2020**, *32*, 1–12. [CrossRef]
13. *Organic Electrochemistry*, 4th ed; Hammerich, O.; Lund, H. (Eds.) Marcel Dekker: New York, NY, USA, 2001.
14. Horvai, G.; Pungor, E. Electrochemical Detectors in HPLC and Ion Chromatography. *Crit. Rev. Anal. Chem.* **1989**, *21*, 1–28. [CrossRef]
15. Toth, K.; Stulik, K.; Kutner, W.; Feher, Z.; Lindner, E. Electrochemical Detection in Liquid Flow Analytical Techniques: Characterization and Classification (IUPAC Technical Report). *Pure Appl. Chem.* **2004**, *76*, 1119–1138. [CrossRef]
16. Xiao, T.F.; Wu, F.; Hao, J.; Zhang, M.N.; Yu, P.; Mao, L.Q. In Vivo Analysis with Electrochemical Sensors and Biosensors. *Anal. Chem.* **2017**, *89*, 300–313. [CrossRef] [PubMed]
17. Wisniewski, N.; Moussy, F.; Reichert, W.M. Characterization of Implantable Biosensor Membrane Biofouling. *Fres. J. Anal. Chem.* **2000**, *366*, 611–621. [CrossRef] [PubMed]
18. Wisniewski, N.; Reichert, M. Methods for Reducing Biosensor Membrane Biofouling. *Colloid Surf. B* **2000**, *18*, 197–219. [CrossRef]
19. Amemiya, S.; Kim, J.; Izadyar, A.; Kabagambe, B.; Shen, M.; Ishimatsu, R. Electrochemical Sensing and Imaging Based on Ion Transfer at Liquid/Liquid Interfaces. *Electrochim. Acta* **2013**, *110*, 836–845. [CrossRef]
20. Izadyar, A. Stripping Voltammetry at the Interface Between Two Immiscible Electrolyte Solutions: A Review Paper. *Electroanalysis* **2018**, *30*, 2210–2221. [CrossRef]
21. Izadyar, A.; Kim, Y.; Ward, M.M.; Amemiya, S. Double-Polymer-Modified Pencil Lead for Stripping Voltammetry of Perchlorate in Drinking Water. *J. Chem. Ed.* **2012**, *89*, 1323–1326. [CrossRef]
22. Kim, Y.; Rodgers, P.J.; Ishimatsu, R.; Amemiya, S. Subnanomolar Ion Detection by Stripping Voltammetry with Solid-Supported Thin Polymeric Membrane. *Anal. Chem.* **2009**, *81*, 7262–7270. [CrossRef]
23. Kim, Y.; Amemiya, S. Stripping analysis of nanomolar perchlorate in drinking water with a voltammetric ion-selective electrode based on thin-layer liquid membrane. *Anal. Chem.* **2008**, *80*, 6056–6065. [CrossRef] [PubMed]
24. Kabagambe, B.; Izadyar, A.; Amemiya, S. Stripping Voltammetry of Nanomolar Potassium and Ammonium Ions Using a Valinomycin-Doped Double-Polymer Electrode. *Anal. Chem.* **2012**, *84*, 7979–7986. [CrossRef] [PubMed]
25. Guo, J.D.; Yuan, Y.; Amemiya, S. Voltammetric Detection of Heparin at Polarized Blood Plasma/1,2-Dichloroethane Interfaces. *Anal. Chem.* **2005**, *77*, 5711–5719. [CrossRef] [PubMed]
26. Amemiya, S.; Kim, Y.; Ishimatsu, R.; Kabagambe, B. Electrochemical Heparin Sensing at Liquid/Liquid Interfaces and Polymeric Membranes. *Anal. Bioanal. Chem.* **2011**, *399*, 571–579. [CrossRef]
27. Amemiya, S.; Yang, X.; Wazenegger, L. Voltammetry of the Phase Transfer of Polypeptide Protamines across Polarized Liquid/Liquid Interfaces. *J. Am. Chem. Soc.* **2003**, *125*, 11832–11833. [CrossRef]

28. Suárez-Herrera, M.F.; Scanlon, M.D. Quantitative Analysis of Redox-Inactive Ions by AC Voltammetry at a Polarised Interface between Two Immiscible Electrolyte Solutions. *Anal. Chem.* **2020**. [CrossRef]
29. Amemiya, S. Voltammetric Ion Selectivity of Thin Ionophore-Based Polymeric Membranes: Kinetic Effect of Ion Hydrophilicity. *Anal. Chem.* **2016**, *88*, 8893–8901. [CrossRef]
30. Kabagambe, B.; Garada, M.B.; Ishimatsu, R.; Amemiya, S. Subnanomolar Detection Limit of Stripping Voltammetric Ca2+-Selective Electrode: Effects of Analyte Charge and Sample Contamination. *Anal. Chem.* **2014**, *86*, 7939–7946. [CrossRef]
31. Kivlehan, F.; Garay, F.; Guo, J.D.; Chaum, E.; Lindner, E. Toward Feedback-Controlled Anesthesia: Voltamrnetric Measurement of Propofol (2,6-Diisopropylphenol) in Serum-Like Electrolyte Solutions. *Anal. Chem.* **2012**, *84*, 7670–7676. [CrossRef]
32. Langmaier, J.; Garay, F.; Kivlehan, F.; Chaum, E.; Lindner, E. Electrochemical quantification of 2,6-diisopropylphenol (propofol). *Anal. Chim. Acta* **2011**, *704*, 63–67. [CrossRef]
33. Srivastava, A.K.; Upadhyay, S.S.; Rawool, C.R.; Punde, N.S.; Rajpurohit, A.S. Voltammetric Techniques for the Analysis of Drugs using Nanomaterials based Chemically Modified Electrodes. *Curr. Anal. Chem.* **2019**, *15*, 249–276. [CrossRef]
34. Pysarevska, S.; Plotycya, S.; Dubenska, L. Voltammetry of Local Anesthetics: Theoretical and Practical Aspects. *Crit. Rev. Anal. Chem.* **2020**. [CrossRef]
35. Amatore, C.; Da Mota, N.; Lemmer, C.; Pebay, C.C.S.; Thouin, L. Theory and Experiments of Transport at Channel Microband Electrodes under Laminar Flows. 2. Electrochemical Regimes at Double Microband Assemblies under Steady State. *Anal. Chem.* **2008**, *80*, 9483–9490. [CrossRef]
36. Bodor, S.; Zook, J.M.; Lindner, E.; Toth, K.; Gyurcsanyi, R.E. Chronopotentiometric Method for the Assessment of Ionophore Diffusion Coefficients in Solvent Ppolymeric Membranes. *J. Solid State Electrochem.* **2009**, *13*, 171–179. [CrossRef]
37. Sheppard, J.B. *Voltammetric Determination of Diffusion and Partition Coefficients in Plasticized Polymer Membranes*; University of Memphis: Memphis, TN, USA, 2016.
38. Takacs-Novak, K.; Avdeef, A. Interlaboratory Study of log P Determination by Shake-Flask and Potentiometric Methods. *J. Pharm. Biomed.* **1996**, *14*, 1405–1413. [CrossRef]
39. Sheppard, J.B.; Hambly, B.; Pendley, B.; Lindner, E. Voltammetric Determination Of Diffusion Coefficients In Polymer Membranes. *Analyst* **2017**, *142*, 930–937. [CrossRef]
40. Rainey, F.; Kivlehan, F.; Chaum, E.; Lindner, E. Toward Feedback Controlled Anesthesia: Automated Flow Analytical System for Electrochemical Monitoring of Propofol in Serum. *Electroanalysis* **2014**, *26*, 1–9. [CrossRef]
41. Kivlehan, F.; Chaum, E.; Lindner, E. Propofol Detection and Quantification in Human Bblood: The Promise of Feedback Controlled, Closed-Loop Anesthesia. *Analyst* **2015**, *140*, 98–106. [CrossRef]
42. Pyka, A.; Babuåka, M.; Zachariasz, M. A Comparison of Theoretical Methods of Calculation of Partition Coefficients for Selected Drugs. *Acta Pol. Pharm.* **2006**, *63*, 159–167.

© 2020 by the authors. Licensee MDPI, Basel, Switzerland. This article is an open access article distributed under the terms and conditions of the Creative Commons Attribution (CC BY) license (http://creativecommons.org/licenses/by/4.0/).

Article

Measurement of Multi Ion Transport through Human Bronchial Epithelial Cell Line Provides an Insight into the Mechanism of Defective Water Transport in Cystic Fibrosis

Miroslaw Zajac [1], Andrzej Lewenstam [2,3], Piotr Bednarczyk [1] and Krzysztof Dolowy [1,*]

1. Institute of Biology, Department of Physics and Biophysics, Warsaw University of Life Sciences—SGGW, 159 Nowoursynowska St., 02-776 Warsaw, Poland; Miroslaw_Zajac@sggw.pl (M.Z.); piotr_bednarczyk@sggw.pl (P.B.)
2. Faculty of Materials Science and Ceramics, AGH University of Science and Technology, Mickiewicza 30, 30-059 Krakow, Poland; andrzej.lewenstam@gmail.com
3. Faculty of Chemistry, Biological and Chemical Research Centre, University of Warsaw, Żwirki i Wigury 101, 02-089 Warsaw, Poland
* Correspondence: krzysztof_dolowy@sggw.pl; Tel.: +48-225938610

Received: 23 January 2020; Accepted: 9 March 2020; Published: 12 March 2020

Abstract: We measured concentration changes of sodium, potassium, chloride ions, pH and the transepithelial potential difference by means of ion-selective electrodes, which were placed on both sides of a human bronchial epithelial 16HBE14σ cell line grown on a porous support in the presence of ion channel blockers. We found that, in the isosmotic transepithelial concentration gradient of either sodium or chloride ions, there is an electroneutral transport of the isosmotic solution of sodium chloride in both directions across the cell monolayer. The transepithelial potential difference is below 3 mV. Potassium and pH change plays a minor role in ion transport. Based on our measurements, we hypothesize that in a healthy bronchial epithelium, there is a dynamic balance between water absorption and secretion. Water absorption is caused by the action of two exchangers, Na/H and Cl/HCO$_3$, secreting weakly dissociated carbonic acid in exchange for well dissociated NaCl and water. The water secretion phase is triggered by an apical low volume-dependent factor opening the Cystic Fibrosis Transmembrane Regulator CFTR channel and secreting anions that are accompanied by paracellular sodium and water transport.

Keywords: ion transport; water transport; epithelium; cystic fibrosis

1. Introduction

Epithelial tissue is made of a single layer of cells bound together by tight junctions. It forms a barrier to the simple diffusion of water and ions. Water is transported across the epithelial cell layer by means of osmosis. Electroneutral transport of a single cation accompanied by a single anion causes the passive osmotic flow of 370 water molecules across the epithelium. In cystic fibrosis (CF), the anion transport route is impaired, which leads to defective ion and water transport across the epithelium and a too dense secreted mucus.

Epithelial cells are polarized, i.e., there are different transporting molecules on both the apical and basolateral surfaces of the epithelial cells. There is a considerable number of different ion channels, transporters and pumps molecules that are present in epithelial cell membranes [1–7].

Despite half a century of research, our knowledge of the interdependence of the ion fluxes and water transport across the epithelial cell layer is limited for several reasons. Firstly, there are many ion channels, transporters and pumps that are involved in transport. Secondly, there are at least five ions transported simultaneously in both directions across the epithelial cell layer i.e., chloride, bicarbonate, sodium, potassium and hydrogen. Thirdly, a method is lacking that would allow for simultaneous measurements of all ions transported within very short time intervals (second-time range). A review of electrochemical and radiotracer methods of ion transport across the epithelium was published recently [8]. There are continuous attempts to increase the number of methods to study epithelial transport e.g., the fluorescence method has been employed [9]. However, these contributions do not provide an explanatory vision of the interdependence of ion fluxes and dense mucus formation.

After a decade of tests and experiments, we succeeded in constructing a device that measures the transport of four ions (sodium, potassium, chloride and pH) across the epithelial cell monolayer grown on porous support by means of ion-selective electrodes [10–12]. Before we started our experiments, it was generally accepted that water moves across the epithelial layer due to the osmotic difference between the apical and basolateral face. Thus, we used two isosmotic solutions that differed with either sodium or chloride concentrations. It was not clear whether a single cell line could transport ions in both directions or only in one. We used gradients with higher sodium or chloride concentrations either on the apical or on the basolateral sides. It was generally believed that ions are transported through the cell i.e., via a transcellular route. Therefore, we used blockers of sodium and chloride channels and transporters in high concentrations to try to prevent transcellular ion transport. The results of our experiments shown in this paper suggest that the generally accepted views in the field of epithelial transport are often wrong.

The results obtained using this method allow offering a more consistent mechanism of apical surface liquid (ASL) equilibrium than any other offered so far. We could now test the hypothesis proposed herein of a dynamic water balance across the human bronchial epithelium.

2. Materials and Methods

2.1. Apparatus

The measuring system has been described earlier [11,12] and is presented in Figure 1. Briefly, the Costar Snapwell inserts with epithelial cell monolayer are sandwiched between two asymmetric probes containing four ion-selective electrodes (Na, K, pH and Cl), the reference electrode, inlet and outlet tubes. The measuring platform is flat and placed 20–40 μm from the surface of the cell layer. The apical and basolateral chamber volumes are similar and equal to approximately 30 μL i.e., only 2–3 times the volume of the cell layer. The electromotive forces (EMFs) of the ion-selective electrodes were measured versus the reference electrode by means of a 16-channel Lawson Lab EMF interface connected to a PC with appropriate data acquisition software. A syringe pump (SP 260PZ, WPI) was used to exchange the media. We kept the speed of media exchange at 0.3 mL/min to avoid damage to the cell layer. The stability of the cell layer was monitored by the measurement of the transepithelial resistance. We observed a dyed medium exchange under the microscope using a modified Costar Snapwells, where the porous membrane was replaced by a glass slide. We found that the medium flow is laminar. After a few chamber volumes had passed, the color of the medium was uniformly distributed within the chamber, excluding the possibility of the medium composition difference parallel to the cell plane. In our experimental conditions, the complete medium exchange took 24 s. Measurements were made after medium flow stopped. The small thickness of the medium layer excluded the possibility of the unstirred layer effect since the ion concentration differences decay within less than one second. The equilibration of our ISE electrodes took 30–45 s as shown in Figure 2. Both tube outlets were open and placed at the same heights to avoid hydrostatic pressure formation. All experiments were performed at room temperature.

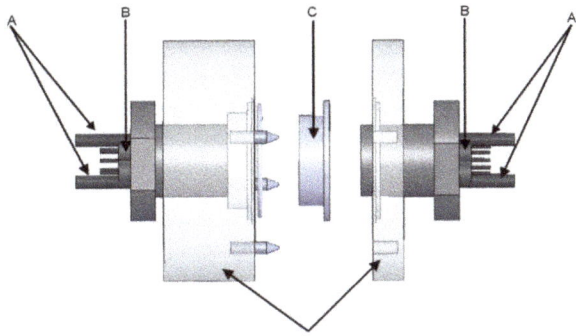

Figure 1. Schematic model of apparatus included. (**A**) solution inlet and outlet tubes, (**B**) ion-selective and reference electrodes, (**C**) Costar Snapwell insert with epithelial cell monolayer, and (**D**) asymmetric body. The diameter of the body is 4 cm. Distance between the cell layer and ion-selective electrodes is less than 30 µm. The inner diameter of the Costar Snapwell insert is 12 mm.

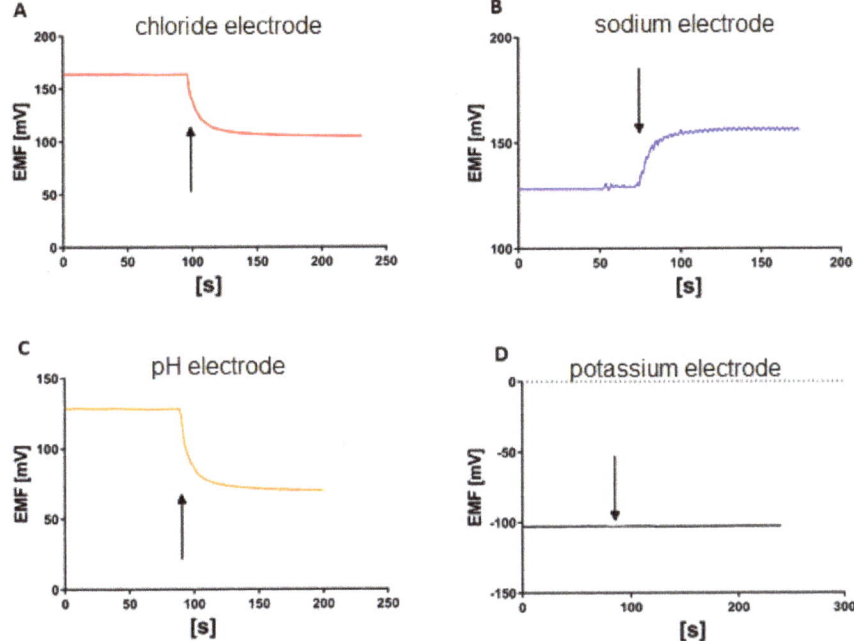

Figure 2. Kinetics of the ion selective electrodes response after changing the medium. The start of medium exchange is shown by the arrow. (**A**) KHS3 solution is replaced by KHS1, (**B**) KHS2 solution is replaced by KHS1, (**C**) KHS1 solution equilibrated to pH = 6.4 is replaced by KHS1 (pH = 7.4), (**D**) KHS3 solution is replaced by KHS1 (no change in potassium concentration).

2.2. Cells

The human bronchial epithelium 16HBE14σ cell line obtained from the late Dr. Dieter Gruenert University of California in San Francisco was grown as described earlier [11,12]. The cells were grown submerged in the culture medium for the first six to nine days. The medium from the apical side was then aspired. Only the basolateral (bottom) side of the cell monolayer was in contact with the medium, while the apical (upper) side was in contact with air supplemented with 5% carbon dioxide. The inserts were used for the experiment after 10–16 days after the air contact was established. The resistance of the cell monolayer used in the experiment was higher than 300 $\Omega \cdot cm^2$. The 16HBEσ cell line is widely used to study respiratory ion transport and the function of the CFTR channel. It is used as the best model of epithelial barrier functions. Additionally, the 16HBEσ cell line model was well suited for our measurements.

2.3. The Media

To study the effect of sodium and chloride concentration gradient across the epithelial cell layer, Krebs–Henseleit solution (KHS1) and its two modifications KHS2 and KHS3 were used (Table 1). All three solutions had the same ionic strength, ionic activity coefficient, pH and osmotic pressure.

Table 1. Ionic composition of media in mM ($\Pi \approx 289$ mOsm/kg H_2O, pH = 7.4).

Ion	KHS1	KHS2	KHS3
Na^+	141.8	24.8	141.8
Cl^-	129.1	129.1	10.0
K^+	5.9	5.9	5.9
Mg^{2+}	1.2	1.2	1.2
Ca^{2+}	2.5	2.5	2.5
HCO_3^-	24.8	24.8	24.8
$H_2PO_4^-$	1.2	1.2	1.2
Choline	-	117.0	-
Gluconate	-	-	119.1
Glucose	11.1	11.1	11.1

To estimate the role of ion channels, transporters and pumps in the transport process, blockers of low specificity and many target molecules in very high doses were used. In the experiments, the KHS solutions were supplemented with 100 µM of amiloride; the potent apical epithelial sodium channel (ENaC) blocker [13] and the blocker of Na/H exchanger, and virtually all sodium-dependent processes; 100 µM of glibenclamide, the apical CFTR channel blocker [14] of K_{ATP} channel and VSOR channel [15] and other exchangers regulated by the ABC binding protein, 100 µM of 4,4'-Diisothiocyano-2,2'-stilbenedisulfonic acid (DIDS), the blocker of the ORCC channel [16,17] and other anion channels and exchangers [18].

The following additives to the media were used: step (I), no additives; (II), amiloride on the apical face of the epithelial cell monolayer; (III), as in step (II) plus glibenclamide and DIDS on both faces of the cell layer.

2.4. Electrodes

The method of fabrication of ion-selective electrodes has been described previously [11]. Each of the electrodes was calibrated separately before mounting them on the measuring platform. The calibration was determined by changing the appropriate ion concentration in the medium. The electrodes showed a linear relationship between the potential and log of ion activity. The slope for the chloride-sensitive electrode was equal to s = −58 ± 2 mV/dec., for the potassium-sensitive electrode +55 ± 1 mV/dec., for the sodium +54 ± 1 mV/dec. and for the hydrogen +57 ± 2 mV/dec. The potential of the ion-selective electrodes was measured against the commercial macro-reference Ag/AgCl electrode (InLab, Mettler Toledo). The micro-reference electrodes used in our system differ from the commercial one by ± 1 mV.

After changing the medium, the ISE potential changes and reaches 95% of the final potential value in 5 min. The slope for the chloride-sensitive electrode was equal to s = −57 ± 2 mV/dec. in KHS1 solution, s = −58 ± 2 mV/dec. in KHS2 and s = −40 ± 2 mV/dec. in KHS3. For the sodium-selective electrode the slope was equal to s = +51 ± 2 mV/dec. in KHS1 solution, s = +40 ± 2 mV/dec. in KHS2 and s = +38 ± 2 mV/dec. in KHS3 solutions.

2.5. Measurement of Multiple Ion Transport Procedure

The epithelial cells grown on porous support were mounted on the measuring platform. The KHS1 solution was introduced to both sides of the cell monolayer. After 10 min on one side of the cell layer (apical or basolateral), KHS2 or KHS3 solution was introduced. The changes in ISE electrode potentials were recorded for 10 min. In the next experimental step, the same solutions supplemented with blockers were introduced to the measuring chambers and the ISE potentials. Measurements were recorded for another 10 min after the medium exchange stop.

3. Results

We performed 269 single determinations measured in 51 series for four different pairs of sodium or chloride concentration gradients using 80 inserts. The results for sodium, chloride and potassium transport are expressed as the change of ion concentration on a particular side of the epithelial layer. Final pH and transepithelial potential difference measured against the potential at basolateral medium 10 min after the media change terminates; for four different media gradients are presented in Figures 3–6. In all cases presented, the concentration gradient induces transport of both sodium and chloride ions from the side of higher to the side of lower concentration. Potassium transport plays a minor role in our experimental conditions. The final pH value is 7.15 ± 0.15 for all media combinations, which means that there is no considerable transport of bicarbonate or proton ions, or they are transported concurrently. It is likely that the pH is controlled in bronchial epithelial cells by at least three processes Na/H exchange, Cl/HCO_3 exchange and by CO_2/HCO_3 equilibrium in which CO_2 production, intracellular and extracellular carbonic anhydrases are involved [7,19]. The transepithelial potential value is less than 3 mV.

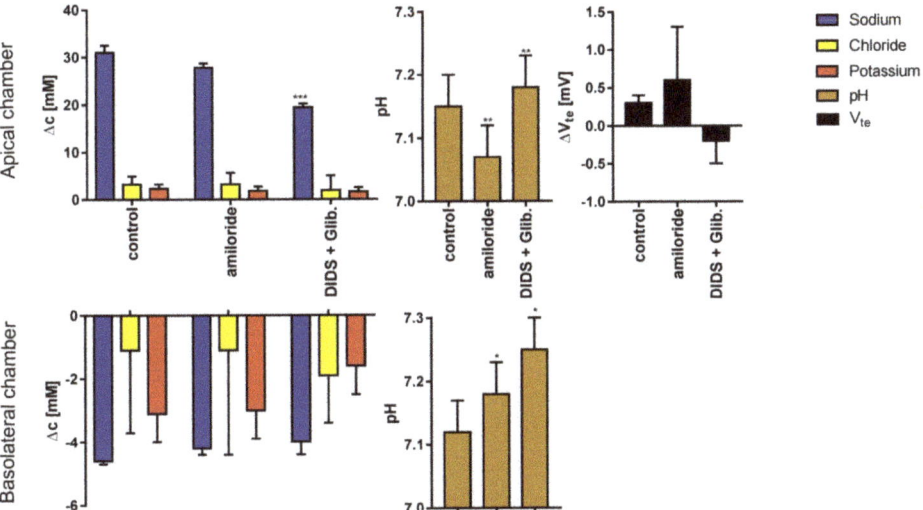

Figure 3. The results (Mean ± SD) of experiments performed in sodium gradient (apical side, KHS2; basolateral side, KHS1) in control conditions and after supplementation of the solution with ion channel blockers. The graph represents: sodium, chloride, potassium transport across the epithelial cell monolayer expressed as the concentration change measured 10 min after appropriate solution flow stop; final pH value and the transepithelial potential difference change (N = 8–10 for Cl^-, Na^+ and V_{te}, N = 6 for K^+ and pH, dependent sample t-test, * $p < 0.05$, ** $p < 0.01$, *** $p < 0.001$). Note different scales for apical and basolateral graphs.

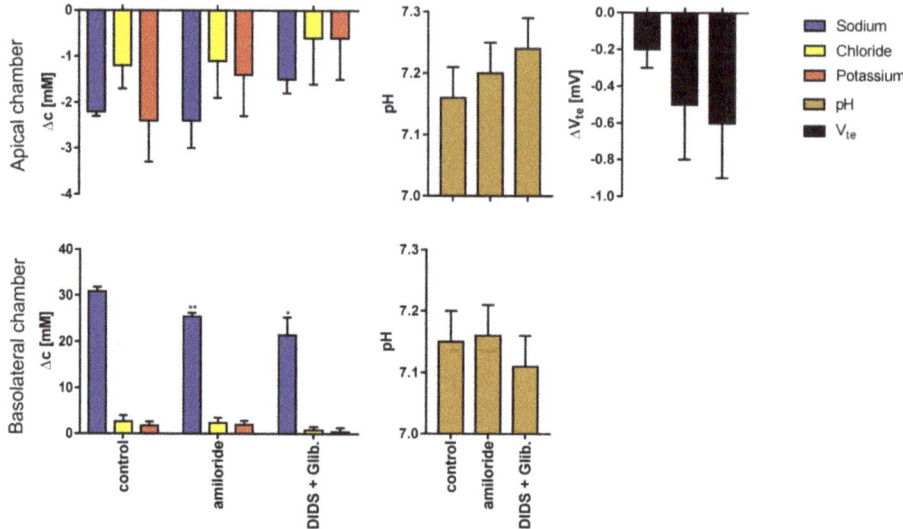

Figure 4. The results (Mean ± SD) of experiments performed in reversed sodium gradient (apical side, KHS1; basolateral side, KHS2) in control conditions and after supplementation of the solution with ion channel blockers. The graph represents: sodium, chloride, potassium transport across epithelial cell monolayer expressed as the concentration change measured 10 min after appropriate solution flow stop; final pH value and the transepithelial potential difference change (N = 7–12 for Cl^-, Na^+ and V_{te}, N = 6 for K^+ and pH, dependent sample t-test, * $p <0.05$, ** $p <0.01$, *** $p <0.001$). Note different scales for apical and basolateral graphs.

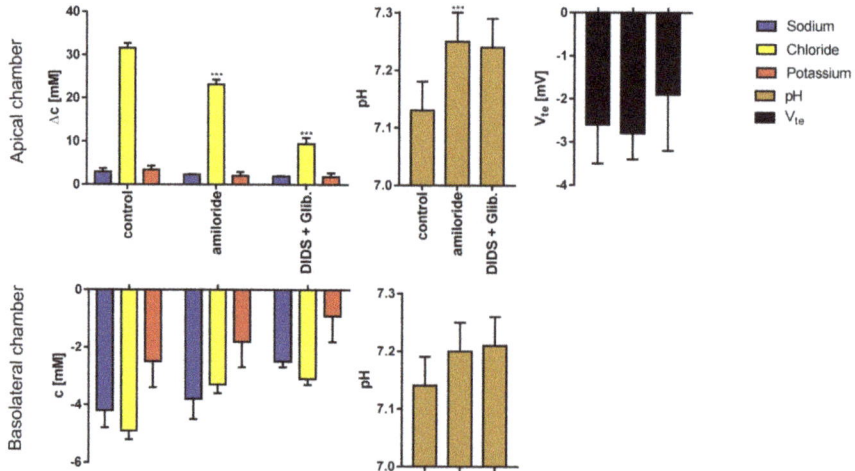

Figure 5. The results (Mean ± SD) of experiments performed in chloride gradient (apical side, KHS3; basolateral side, KHS1) in control conditions and after supplementation of the solution with ion channel blockers. The graph represents: sodium, chloride and potassium transport across epithelial cell monolayer expressed as the concentration change measured 10 min after appropriate solution flow stop; final pH value and the transepithelial potential difference change (N = 15–17 for Cl^-, Na^+ and V_{te}, N = 5 for K^+ and pH, dependent sample t-test, * $p <0.05$, ** $p <0.01$, *** $p <0.001$). Note different scales for apical and basolateral graphs.

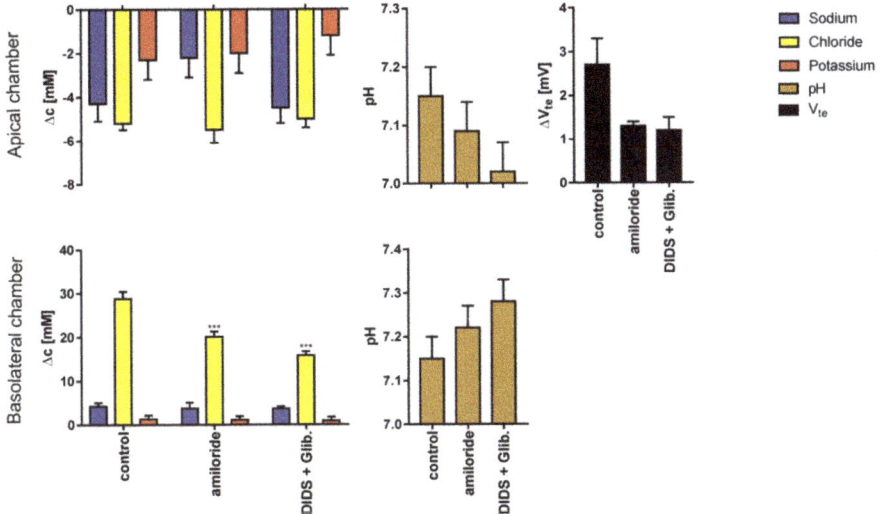

Figure 6. The results (Mean ± SD) of experiments performed in reversed chloride gradient (apical side, KHS1; basolateral side, KHS3) in control conditions and after supplementation of the solution with ion channel blockers. The graph represents: sodium, chloride, potassium transport across epithelial cell monolayer expressed as the concentration change measured 10 min after appropriate solution flow stop; final pH value and the transepithelial potential difference change (N = 10–13 for Cl^-, Na^+ and V_{te}, N = 5 for K^+ and pH, dependent sample t-test, * $p < 0.05$, ** $p < 0.01$, *** $p < 0.001$). Note different scales for apical and basolateral graphs.

4. Discussion

The data presented in Figures 3–6 show that the measured transport of cations is not equal to anions i.e., the transport is apparently not electroneutral. The transport of ions to the chamber with lower concentrations of chloride and sodium is of one order of magnitude higher than the decrement on the higher concentration side. The simple explanation that the export or import of bicarbonate anion provides the balance does not hold since bicarbonate transport should also affect the pH and the required amount of bicarbonate surpasses the amount available in the chambers.

Thus, another explanation for the apparent breaking of the electroneutrality and mass transport principle is required. We recalculated the concentration changes to the final concentration values which show that the concentration of the anions is equal to cation concentration—thus the electroneutrality principle is not violated. This allows us to assume that the change in concentrations of ions is caused by isosmotic transport of 145 mM NaCl solution across the cell layer and that the transported solution consists only of x fraction of the chamber volume. From this assumption, one can calculate the expected final concentration of particular ion C_{fin} starting from the initial concentration C_{init} as $145x + C_{init} * (1 - x) = C_{fin}$ for the side to which NaCl solution is transported and $C_{init} * (1 + x) - 145x = C_{fin}$ for the side from which NaCl solution is transported. The volume of transported fluid (x) is in control conditions, approximately 25% of the chamber volume (equivalent to 5–10 µL) and less for the solutions supplemented with (Table 2). Using x values from Table 2, one obtains an almost perfect fit between the theoretical and the measured values of final sodium and chloride concentrations (Figures 7 and 8).

We found that the imbalance of sodium or chloride ions concentrations on both sides of the epithelial cell monolayer causes the flux of isosmotic NaCl solution across the epithelial layer and thus decreasing the ion gradient. We found that the transepithelial resistance of the cell monolayer is not affected during our experimental procedure, which rules out the formation of the permanent fluid transport route across the cell layer. The results presented in Figures 7 and 8 and Table 2 suggest that the blockers of chloride ion channels and transporters considerably affect the volume of fluid transported, while the blockers of sodium channels and transporters are less effective. This is in accordance with earlier finding that sodium is transported via a paracellular way while chloride is transported via a transcellular one [11]. According to Table 2, during 600 s of the experiment, 25% of 30 µL of chamber volume is transported across the 1.1 cm^2 cell layer in the form of 145 mM NaCl solution; this is equivalent to the movement of 10^{15} ions per 1.1 cm^2 per second or 1.5 pA/µm^2 which can be delivered by a few open CFTR channels per µm^2 or a single VSOR or ORCC channels per few µm^2. However, further studies of the mechanism of fluid transport are necessary e.g., on whether the water is transported transcellularly via aquaporins [20] or paracellularly. While our experiments considered drastic nonequilibrium conditions, similar mechanisms are likely to be involved in near-equilibrium conditions in the lungs.

Table 2. Fractions of chamber volume transported in experimental conditions calculated to obtain the best fit between experimental and theoretical predictions.

Buffer in the Chamber		Fraction of Transported Chamber Volume Fluid (%)		
Apical	Basolateral	Control	Amiloride	DIDS+Glibenclamide
KHS2	KHS1	25.9	23.1	17.1
KHS1	KHS2	25.0	22.5	18.6
KHS3	KHS1	24.2	17.5	7.9
KHS1	KHS3	22.5	16.9	14.1

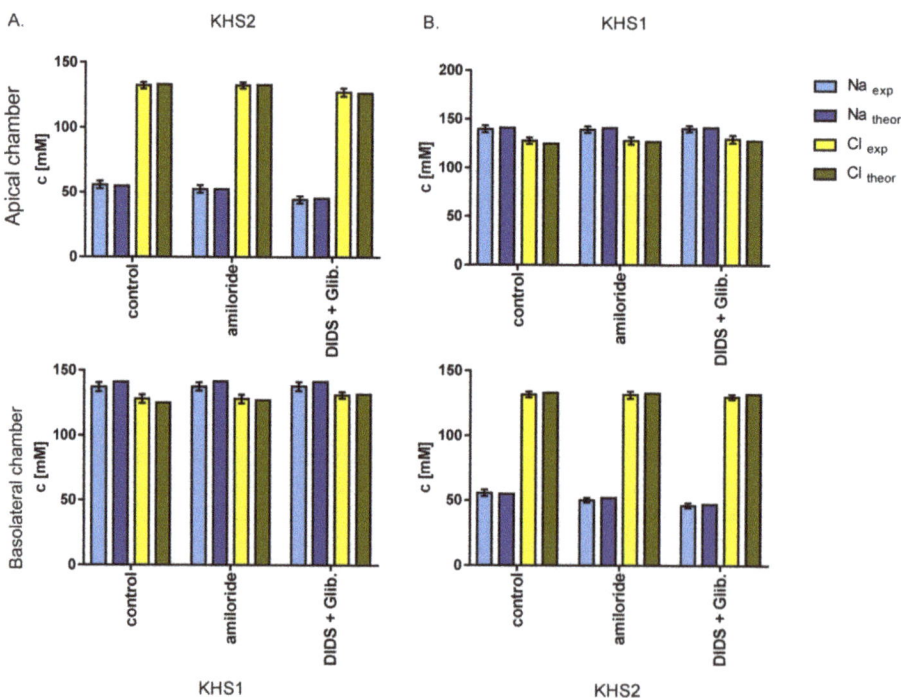

Figure 7. Measured (Mean ± SD) and theoretically predicted concentrations of sodium and chloride ions on both sides of the cell layer in control conditions after blocking particular ion channels present on the apical side of the monolayer. The experiments with sodium gradients and isosmotic flow of x fraction of 145 mM sodium chloride across the epithelial cell monolayer are shown.

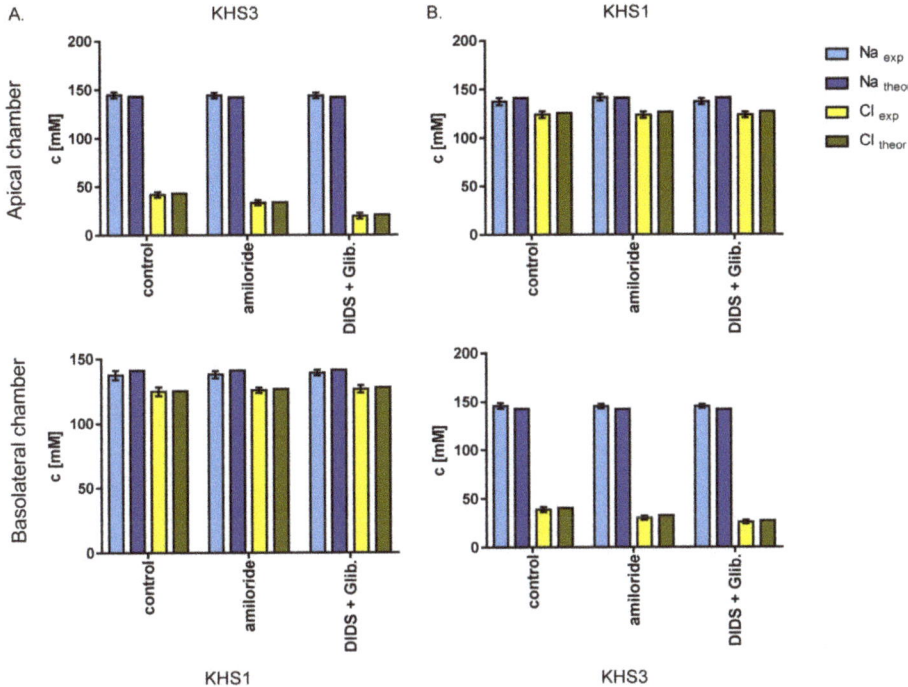

Figure 8. Measured (Mean ± SD) and theoretically predicted concentrations of sodium and chloride ions on both sides of the cell layer in control conditions after blocking particular ion channels present on the apical side of the monolayer. The experiments with sodium gradients and isosmotic flow of x fraction of 145 mM sodium chloride across the epithelial cell monolayer are shown.

There are a few hypotheses of how epithelial cells regulate the thickness of apical fluid in vivo. Boucher [21,22] suggests that the same cells alternate between fluid secretion and absorption. Quinton [23] claims there are separate cells that independently secrete and absorb fluid. There is strong evidence that bicarbonate ions are transported across epithelial cells and are responsible for proper mucus consistency [24,25]. Considering our experimental results presented herein show that: (i) water is transported in both directions, (ii) bicarbonate and proton ions are transported concurrently, and (iii) our earlier findings that sodium ions are transported paracellularly while chloride ions are transported transcellularly [11], we propose 'the dynamic water balance' hypothesis and relevant model. The essence of our concept is presented in Figure 9. We assume that in a healthy bronchial epithelium, there is a dynamic balance between water absorption (Figure 9A) and water secretion (Figure 9B). The action of two exchangers, Na/H and Cl/HCO$_3$, is responsible for the water absorption phase. In effect, with the ion transport conducted by these two exchangers, there is a flux of sodium and chloride into the cytoplasm and secretion of bicarbonate and proton ions from the cytoplasm to apical fluid (ASL). While sodium chloride is fully dissociated, the bicarbonate and proton form weakly dissociated carbonic acid, decreasing osmotic pressure. The lower osmotic pressure of ASL causes water transport into the cytoplasm; also, the carbonic acid present in ASL fluid slowly decomposes due to the presence of a small amount of carbonic anhydrase in the apical fluid [26]. The observation of Alexandrou and Walters [27] that the absorption of lung fluid is amiloride-sensitive (acting on Na/H exchanger) but not sensitive to chloride channels blocker is in accordance with our model. The exchange of chloride for bicarbonate results in an increase of bicarbonate concentration in the apical fluid and chloride concentration in the cytoplasm (and a decrease on the opposite sides). The energy gain of chloride for bicarbonate exchange is not far from electrochemical equilibrium. The small change of concentration of bicarbonate ion in the cytoplasm by only 5 mM and the increase of 5 mM in the ASL fluid (or 20 mM of chloride concentration change) would stop net ion transport through the Cl/HCO$_3$ exchanger, while the Na/H exchanger will still operate leading to the acidification of ASL fluid.

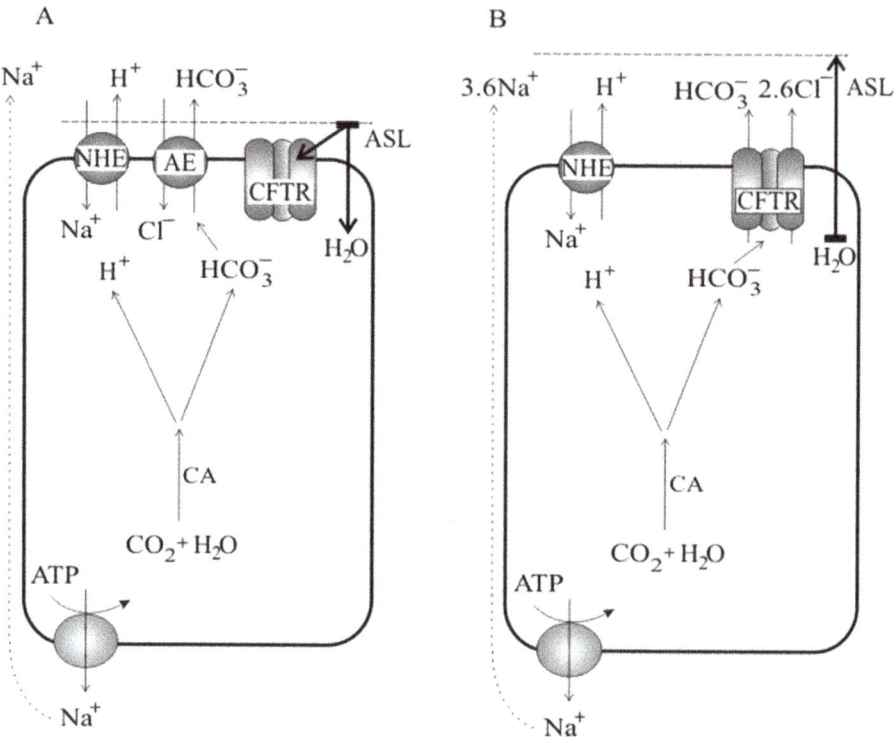

Figure 9. 'The dynamic water balance' hypothesis: (**A**) water absorption, (**B**) water secretion. During the absorption phase (**A**) two exchangers are responsible for sodium and chloride influx from ASL into the cytoplasm and outward proton and bicarbonate secretion. The lower osmotic pressure of partially dissociated carbonic acid in ASL causes water absorption and depletion of its volume, which increases the concentration of the ASL volume-dependent factor (e.g., ATP) triggering the opening of the CFTR channel and starting the secretion phase (**B**). The secretion of chloride and bicarbonate ions via the CFTR channel is accompanied by the transport of sodium ions via a paracellular pathway. The appearance of net 2.6 NaCl molecules on the apical side leads to osmotic water transport, ASL hydration and dilution of the triggering factor resulting in returning the epithelium to the absorption phase. In the absence of functional CFTR, the bronchial epithelium stays in the absorptive phase leading to Cl/HCO$_3$ reaching energetic equilibrium, the action of Na/H exchanger, acidification of ASL and undiluted mucus.

During the absorptive phase, the volume of ASL decreases. We assume that an ASL volume-dependent factor i.e., ATP concentration in ASL as proposed earlier [28,29], triggers the opening of the CFTR channel. The opening of the CFTR channel starts the secretion phase (Figure 9B). The secretion of chloride and bicarbonate ions through the CFTR channel is accompanied by paracellular transport of sodium ions and the action of the Na/H exchanger. Taking the ion selectivity ratio of $k_{Cl/HCO3}$ = 4 [30,31] for the CFTR channel, one can easily estimate that 2.6 chloride ion is secreted to ASL for 1 bicarbonate:

$$\frac{I_{Cl}}{I_{HCO3}} = \frac{k_{Cl/HCO3} \times \left(\frac{RT}{F} \ln\left(\frac{[Cl_c]}{Cl_{ASL}}\right) + (\varphi_{ASL} - \varphi_c)\right)}{\frac{RT}{F} \times \ln\left(\frac{[HCO3_c]}{[HCO3_{ASL}]}\right) + (\varphi_{ASL} - \varphi_c)} = 2.6$$

where subscript 'ASL' denotes apical fluid and subscript 'c' cytoplasm; during the secretory phase, the net 2.6 molecules of NaCl appear on the apical side of the bronchial epithelium. The osmotic pressure of ASL increases, leading to water secretion to ASL. Water secretion dilutes the trigger molecule responsible for opening the CFTR channel. The CFTR channel closes, and the system switches back to the absorptive phase. In the absence of a functional CFTR channel, the trigger sensor is missing, and the secretory phase is no longer self-regulated by the end of the absorptive phase. The consequence of the lack of a functional CFTR channel is the low volume of ASL, dense mucus layer, and the acidification of ASL characteristic of CF. In this way, the hypothesis presented combines ion fluxes with water transport and explains the formation of ASL mucus in the case of dysfunctional CFTR.

The secretion of chloride into ASL is possible also via other anionic channels present in the apical face of the bronchial epithelium e.g., by the calcium-dependent anionic channel CaCC modeled recently [32]. However, the secretion through the CaCC channel (TMEM16A) due to the lower selectivity ratio $k_{Cl/HCO3}$ = 2 [33], or even less [34], would lead to secretion of only 1.3 chloride per 1 bicarbonate ion and consequently, only half of the amount of water is secreted compared to a functional CFTR channel. Moreover, the system is not self-regulated.

5. Conclusion

In the isosmotic transepithelial concentration gradient of either sodium or chloride ions, there is an electroneutral transport of the isosmotic solution of sodium chloride in both directions across the cell monolayer.

In a healthy bronchial epithelium, there is a dynamic balance between water absorption and secretion. Water absorption is caused by the action of two exchangers, Na/H and Cl/HCO$_3$, secreting weakly dissociated carbonic acid in exchange for well dissociated NaCl and water. The water secretion phase is triggered by an apical low volume-dependent factor opening the CFTR channel and secreting anions that are accompanied by paracellular sodium and water transport.

Author Contributions: Conceptualization, A.L. and K.D.; methodology, M.Z., A.L. and K.D.; validation, P.D. and K.D.; investigation, M.Z.; data curation, M.Z.; writing—original draft preparation, K.D.; writing—review and editing, M.Z., A.L. and P.B.; visualization, M.Z. and K.D.; supervision, K.D. All authors have read and agreed to the published version of the manuscript.

Funding: Andrzej Lewenstam was supported by research grant no. 2014/15/B/ST5/02185, Miroslaw Zajac by research grant no. 2018/02/X/NZ4/00304, Piotr Bednarczyk by research grant no. 2016/21/B/NZ1/02769 all from the National Science Centre (NCN, Poland). The research was partially funded by Wlasny Fundusz Stypendialny SGGW, granted for Miroslaw Zajac.

Acknowledgments: Drawing of Figure 9 by Ms. Wanda Jarzabek is appreciated.

Conflicts of Interest: The authors declare no conflict of interest.

References

1. Hollenhorst, M.I.; Richter, K.; Fronius, M. Ion Transport by Pulmonary Epithelia. *J. Biomed. Biotechnol.* **2011**, *2011*, 174306. [CrossRef] [PubMed]
2. Toczylowska-Maminska, R.; Dolowy, K. Ion transporting proteins of human bronchial epithelium. *J. Cell. Biochem.* **2012**, *113*, 426–432. [CrossRef] [PubMed]
3. Reichhart, N.; Strauss, O. Ion channels and transporters of the retinal pigment epithelium. *Exp. Eye Res.* **2014**, *126*, 27–37. [CrossRef] [PubMed]
4. Mall, M.A.; Galietta, L.J.V. Targeting ion channels in cystic fibrosis. *J. Cyst. Fibros.* **2015**, *14*, 561–570. [CrossRef]
5. Bartoszewski, R.; Matalon, S.; Collawn, J.F. Ion channels of the lung and their role in disease pathogenesis. *Am. J. Physiol. Lung Cell. Mol. Physiol.* **2017**, *313*, L859–L872. [CrossRef] [PubMed]

6. Novak, I.; Praetorius, J. Fundamentals of Bicarbonate Secretion in Epithelia. In *Ion Channels and Transporters of Epithelia in Health and Disease. Physiology in Health and Disease*; Hamilton, K., Devor, D., Eds.; Springer: New York, NY, USA, 2016; pp. 187–263. [CrossRef]
7. Webster, M.J.; Tarran, R. Slippery When Wet: Airway Surface Liquid Homeostasis and Mucus Hydration. *Cell Vol. Regul.* **2018**, *81*, 293–335. [CrossRef]
8. Zajac, M.; Dolowy, K. Measurement of ion fluxes across epithelia. *Prog. Biophys. Mol. Biol.* **2017**, *127*, 1–11. [CrossRef]
9. Munkonge, F.; Alton, E.W.; Andersson, C.; Davidson, H.; Dragomir, A.; Edelman, A.; Farley, R.; Hjelte, L.; McLachlan, G.; Stern, M.; et al. Measurement of halide efflux from cultured and primary airway epithelial cells using fluorescence indicators. *J. Cyst. Fibros.* **2004**, *3*, 171–176. [CrossRef]
10. Toczylowska-Maminska, R.; Lewenstam, A.; Dolowy, K. Multielectrode Bisensor System for Time-Resolved Monitoring of Ion Transport Across an Epithelial Cell Layer. *Anal. Chem.* **2014**, *86*, 390–394. [CrossRef]
11. Zajac, M.; Lewenstam, A.; Dolowy, K. Multi-electrode system for measurement of transmembrane ion-fluxes through living epithelial cells. *Bioelectrochemistry* **2017**, *117*, 65–73. [CrossRef]
12. Zajac, M.; Lewenstam, A.; Stobiecka, M.; Dolowy, K. New ISE-based apparatus for Na^+, K^+, Cl^-, pH and transepithelial potential difference real-time simultaneous measurements of ion transport across epithelial cells monolayer-advantages and pitfalls. *Sensors* **2019**, *19*, 1881. [CrossRef] [PubMed]
13. Anagnostopoulou, P.; Dai, L.; Schatterny, J.; Hirtz, S.; Duerr, J.; Mall, M.A. Allergic airway inflammation induces a pro-secretory epithelial ion transport phenotype in mice. *Eur. Respir. J.* **2010**, *36*, 1436–1447. [CrossRef] [PubMed]
14. Schultz, B.D.; DeRoos, A.D.G.; Venglarik, C.J.; Singh, A.K.; Frizzell, R.A.; Bridges, R.J. Glibenclamide blockade of CFTR chloride channels. *Am. J. Physiol. Lung Cell. Mol. Physiol.* **1996**, *271*, L192–L200. [CrossRef]
15. Okada, Y.; Sato, K.; Numata, T. Pathophysiology and puzzles of the volume-sensitive outwardly rectifying anion channel. *J. Physiol.* **2009**, *587*, 2141–2149. [CrossRef] [PubMed]
16. Hwang, T.; Lee, H.; Lee, N.; Choi, Y.C. Evidence that basolateral but not apical membrane localization of outwardly rectifying depolarization induced Cl channel in airway epithelia. *J. Membr. Biol.* **2000**, *176*, 217–221. [CrossRef] [PubMed]
17. Szkotak, A.J.; Man, S.F.P.; Duszyk, M. The role of basolateral outwardly rectifying chloride channel in human airway epithelial anion secretion. *Am. J. Respir. Cell Mol. Biol.* **2003**, *29*, 710–720. [CrossRef]
18. Blaisdell, C.J.; Edmonds, R.D.; Wang, X.T.; Guggino, S.; Zeitlin, P.L. pH-regulated chloride secretion in fetal lung epithelia. *Am. J. Physiol. Lung Cell. Mol. Physiol.* **2000**, *278*, L1248–L1255. [CrossRef]
19. Tresguerres, M.; Buck, J.; Levin, L.R. Physiological carbon dioxide, bicarbonate, and pH sensing. *Pflügers Arch.-Eur. J. Physiol.* **2010**, *460*, 953–964. [CrossRef]
20. Sharma, S.; Kumaran, G.K.; Hanukoglu, I. High-resolution imaging af the actin cytoskeleton and epithelial sodium channel, CFTR, and aquaporin-9 localization in the vas deferens. *Mol. Reprod. Dev.* **2020**, 1–15. [CrossRef]
21. Boucher, R.C. Human airway ion transport. Part one. *Am. J. Respir. Crit. Care Med.* **1994**, *150*, 271–281. [CrossRef]
22. Boucher, R.C. Human airway ion transport. Part two. *Am. J. Respir. Crit. Care Med.* **1994**, *150*, 281–293. [CrossRef] [PubMed]
23. Quinton, P.M. Both ways at once: Keeping small airways clean. *Physiology* **2017**, *32*, 380–390. [CrossRef] [PubMed]
24. Shamsuddin, A.K.; Quinton, P.M. Concurrent absorption and secretion of airway surface liquids and bicarbonate secretion in human brachioles. *Am. J. Physiol. Lung Cell. Mol. Physiol.* **2019**, *316*, L953–L960. [CrossRef] [PubMed]
25. Kunzelmann, K.; Schreiber, R.; Hadorn, H.B. Bicarbonate in cystic fibrosis. *J. Cyst. Fibros.* **2017**, *16*, 653–662. [CrossRef]
26. Thornell, I.M.; Li, X.P.; Tang, X.X.; Brommel, C.M.; Karp, P.H.; Welsh, M.J.; Zabner, J. Nominal carbonic anhydrase activity minimizes airway-surface liquid pH changes during breathing. *Physiol. Rep.* **2018**, *6*. [CrossRef]
27. Alexandrou, D.; Walters, D.V. The role of Cl− in the regulation of ion and liquid transport in the intact alveolus during β-adrenergic stimulation. *J. Exp. Physiol.* **2013**, *98*, 576–584. [CrossRef]

28. Cantiello, H.F.; Prat, A.G.; Reisin, I.L.; Ercole, L.B.; Abraham, E.H.; Amara, J.F.; Gregory, R.J.; Ausiello, D.A. External ATP and its analogs activate the cystic-fibrosis transmembrane conductance regulator by a cyclic amp-independent mechanism. *J. Biol. Chem.* **1994**, *269*, 11224–11232.
29. Lazarowski, E.R.; Tarran, R.; Grubb, B.R.; van Heusden, C.A.; Okada, S.; Boucher, R.C. Nucleotide release provides a mechanism for airway surface liquid homeostasis. *J. Biol. Chem.* **2004**, *279*, 36855–36864. [CrossRef]
30. Cai, Z.W.; Chen, J.H.; Hughes, L.K.; Li, H.Y.; Sheppard, D.N. The Physiology and Pharmacology of the CFTR Cl- Channel. In *Chloride Movements across Cellular Membranes*, 1st ed.; Pursch, M., Ed.; Elsevier Science: San Diego, CA, USA, 2007; Volume 38, pp. 109–143. [CrossRef]
31. Tang, L.; Fatehi, M.; Linsdell, P. Mechanism of direct bicarbonate transport by the CFTR anion channel. *J. Cyst. Fibros.* **2009**, *8*, 115–121. [CrossRef]
32. Sandefur, C.I.; Boucher, R.C.; Elston, T.C. Mathematical model reveals role of nucleotide signaling in airway surface liquid homeostasis and its dysregulation in cystic fibrosis. *Proc. Natl. Acad. Sci. USA* **2017**, *114*, E7272–E7281. [CrossRef]
33. Wei, L.; Vankeerberghen, A.; Cuppens, H.; Eggermont, J.; Cassiman, J.J.; Droogmans, G.; Nilius, B. Interaction between calcium-activated chloride channels and the cystic fibrosis transmembrane conductance regulator. *Pflügers Arch.* **1999**, *438*, 635–641. [CrossRef] [PubMed]
34. Jung, J.; Nam, J.H.; Park, H.W.; Oh, U.; Yoon, J.H.; Lee, M.G. Dynamic modulation of ANO1/TMEM16A HCO_3^- permeability by Ca^{2+}/calmodulin. *Proc. Natl. Acad. Sci. USA* **2013**, *110*, 360–365. [CrossRef] [PubMed]

© 2020 by the authors. Licensee MDPI, Basel, Switzerland. This article is an open access article distributed under the terms and conditions of the Creative Commons Attribution (CC BY) license (http://creativecommons.org/licenses/by/4.0/).

Article

Precipitation of Inorganic Salts in Mitochondrial Matrix

Jerzy J. Jasielec [1,*], Robert Filipek [1], Krzysztof Dołowy [2] and Andrzej Lewenstam [1,3]

[1] Faculty of Materials Science and Ceramics, AGH University of Science and Technology, Al. Mickiewicza 30, 30-059 Krakow, Poland; rof@agh.edu.pl
[2] Department of Physics and Biophysics, Warsaw University of Life Sciences—SGGW, Nowoursynowska 159, 02-776 Warszawa, Poland; krzysztof_dolowy@sggw.edu.pl
[3] Johan Gadolin Process Chemistry Centre, Centre of Process Analytical Chemistry and Sensor Technology (ProSens), Åbo Akademi University, Biskopsgatan 8, 20500 Åbo-Turku, Finland; alewenst@abo.fi
* Correspondence: jasielec@agh.edu.pl; Tel.: +48-517-149-455

Received: 26 March 2020; Accepted: 19 April 2020; Published: 27 April 2020

Abstract: In the mitochondrial matrix, there are insoluble, osmotically inactive complexes that maintain a constant pH and calcium concentration. In the present paper, we examine the properties of insoluble calcium and magnesium salts, such as phosphates, carbonates and polyphosphates, which might play this role. We find that non-stoichiometric, magnesium-rich carbonated apatite, with very low crystallinity, precipitates in the matrix under physiological conditions. Precipitated salt acts as pH buffer, and, hence, can contribute in maintaining ATP production in ischemic conditions, which delays irreversible damage to heart and brain cells after stroke.

Keywords: mitochondrion; calcium carbonates; calcium phosphates; calcium polyphosphates; ATP production; hypoxia; ischemia; pre-conditioning

1. Introduction

Mitochondria are responsible for the adenosine triphosphate (ATP) production in eukaryotic organisms. They are also involved in Ca^{2+} signalling, lipid metabolism, heat production, reactive oxygen species production and apoptosis [1]. It is becoming increasingly apparent that mitochondria can accumulate, store and release a large amount of calcium ions under a variety of physiological and pathological conditions such as epilepsy [2], ischemia [3,4] and concussive brain injury [5].

The mitochondrial matrix is surrounded by two membranes. The outer membrane acts as a molecular sieve, while the inner membrane is ion-selective. The major function of the mitochondrial inner membrane is the formation of the electrochemical proton gradient, known as proton motive force [6], utilizing the energy produced by burning fats and sugars. The proton motive force is then used by ATP synthase to synthetize ATP from adenosine diphosphate (ADP) and inorganic phosphate. In the mitochondrial inner membrane there are numerous ion channels and ion exchangers which are involved in the transport of calcium, magnesium, potassium, sodium, protons, phosphates and other ions. The most relevant transporters in the mitochondrial inner membrane are depicted in Scheme 1. Due to the very high electric potential difference (inside negative), the mitochondrial matrix tends to accumulate calcium ions, which form a complex of calcium phosphates and carbonates. Of interest in this report are free calcium and magnesium ions, and the composition of the solid aggregates precipitating in the matrix.

The free concentration of Ca^{2+} and Mg^{2+} depends on complex chemistry which is not easy to decode because of thermodynamic and kinetic factors. The capacity of isolated mitochondria to accumulate calcium up to total matrix concentration of 0.5 M to 0.8 M, has been established more than half a century ago [7–11]. However, the free calcium concentration in the range of 0.17 to 5 µM has been reported by numerous studies using fluorescent Ca^{2+} indicators (see Table 1).

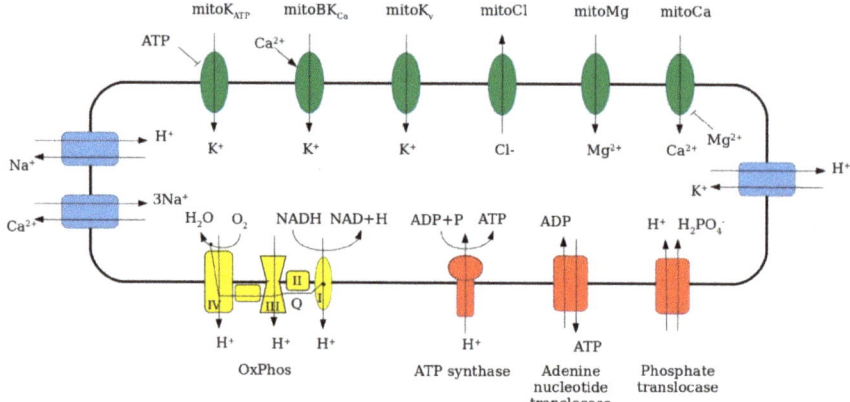

Scheme 1. Ion channels, exchangers, pumps and oxidative phosphorylation chain of the inner mitochondrial membrane: green—uniporters, blue—exchangers, yellow—four complexes of oxidative phosphorylation chain, and red—ATP synthasome. There are more than one type of mitoBK$_{Ca}$, mitoKv or mitoCl channels.

In mitochondria calcium has important physiological role in stimulation of ATP synthesis via activation of several Ca^{2+}-sensitive enzymes (pyruvate-, α-ketoglutarate-, isocitrate dehydrogenases), ATP synthase and adenine nucleotide translocase [12,13]. Matrix enzymes are activated by calcium of concentration in the micromolar range, i.e., $[Ca^{2+}]_{matrix}$ = 0.5 to 2 µM [14]. Calcium can be accumulated in or released from the mitochondrion, depending on the proton motive force of the mitochondrion and cytoplasmic free calcium concentration.

In the presence of phosphates, calcium appears to form an insoluble and osmotically inactive complex, which buffer calcium concentration [7]. Amorphous granules containing calcium and phosphorus were identified within the mitochondria of osteoclasts [15], chondrocytes [16,17], osteoblasts [18], osteocytes [19–21], calcifying cartilage [22], and mineralizing bone [23–25]. It is impossible to investigate the nature of the calcium phosphates in the matrix directly, due to their instant dissociation when mitochondria are disrupted and due to their ability to transform during fixation or drying for chemical analysis [26]. Additionally, it is reasonable to presume that these complexes in the matrix are inert, to some extend at least. Even more importantly, the composition of mitochondrial granules is still largely unknown [27] and it is based on several hypotheses. Thomas and Greenawalt [28] suggested that inorganic granules found in mitochondria are a colloidal, sub-crystalline precursor of calcium-deficient hydroxyapatite. Lehninger [8] and later Nicholls and Chalmers [29] assumed the formation of amorphous tricalcium phosphate, while Kristian et al. [27] suggested the formation of brushite and non-stoichiometric amorphous calcium phosphate with Ca/P ratio equal 1.47. Dołowy, in the most recent hypothesis [30], assumed that brushite turns into isoclasite via all intermediate forms of calcium phosphate minerals forming a polymer-like structure.

The inorganic cations present in the mitochondrial matrix include Na^+, K^+, Ca^{2+} and Mg^{2+}. All sodium and potassium salts are soluble. Calcium and magnesium, on the other hand, can form a variety of insoluble salts. The solubility products of calcium and magnesium hydroxides are relatively high ($K_{Ca(OH)2}$ = 6.5 × 10^{-6} and $K_{Mg(OH)2}$ = 5.6 × 10^{-12}), therefore in considered system $Ca(OH)_2$ and $Mg(OH)_2$ will dissociate completely. In this work, the chemical equilibria of numerous compounds, including calcium and magnesium carbonates, phosphates and polyphosphates are intentionally examined. Based on the solubility products and other properties of these salts, we identify the reason for the pH buffering effect, i.e., the calcium hydroxide buffer effect in the mitochondrial matrix.

2. Calculations and Data

2.1. Mitochondrial pH, Ionic Concentrations and Activities

One must realize the meaning of pH in the systems with very low volumes. The amount of protons can be calculated as $n_{H+} = N_A \times V \times 10^{-pH}$, where $N_A = 6.02 \times 10^{-23}$ is the Avogadro number and V represents volume. The mitochondrion is an organelle with a diameter of c.a. 1 µm, hence the volume of ~1 fL. The mitochondrial matrix occupies 90% of mitochondrial volume. With $pH_{mat} = 8.2$, one obtains 3.4 protons in the matrix. This indicates that water is too weakly dissociated to be a direct donor or acceptor of protons at very low biological volumes (even if one can still define the pH formally) and that protons can be shuttled to reaction centers by other biomolecules, whose concentrations are not constrained by the solubility product of water [31]. These protons may be provided by carbonic or phosphoric acid or dissolution of their salts.

The molecular probes used for pH measurements, like e.g., SNARF-1 (seminaphtharhodafluor) [32–34] or mutants of green fluorescent protein [35–37], report their own protonation state and not the concentration of free protons in the solution.

The values of pH and the concentrations of inorganic ions, measured in mitochondria using various methods, are listed in Table 1. The mitochondrion is a complex organelle in which pH and concentrations of ions fluctuate in response to metabolic processes and perturbations of the cytosolic environment [38]. Mitochondrial pH is reported to be in the range of 7.2 to 8.4 with an average value of 7.8 (see Table 1). However, when mitochondria are loaded with succinate, the pH can reach values close to 9 [39]. The concentration of calcium in mitochondria usually varies between 0.17 to 5 µM (see Table 1). However, in the case of mitochondria of firing axon, it can reach values as high as 24 µM [40]. Furthermore, the concentrations of ions are interdependent, e.g., calcium effluxes from mitochondria can depend on the concentration of sodium [41].

The driving force for the transition of dissolved matter from a supersaturated solution to the equilibrium solid phase, is given by the change of standard free energy $\Delta G^0 = -RT\ln S$, where $S = Q/K_{sp}$ is supersaturation defined as the ratio of the ion activity product (Q) of constituent ions in solution over its equilibrium solubility product (K_{sp}). If $S > 1$, the solution is supersaturated, and precipitation may occur. If $S < 1$, the solution is undersaturated, and there will be no precipitation.

The activity product of a salt $A_x B_y$ is defined as:

$$Q = \{A\}^x \times \{B\}^y = \gamma_A^x \times \gamma_B^y \times [A]^x \times [B]^y \tag{1}$$

where $\{A\}$ and $\{B\}$ are the activities, $[A]$ and $[B]$ are the concentrations, γ_A and γ_B are the activity coefficients of ions A and B, respectively.

In ideal solution, the activity coefficients of all ions equal one. However, in the concentrated solutions γ_i can significantly vary from unity and strongly depends on the ionic strength of the solution defined as:

$$I = 0.5 \sum z_i^2 c_i \tag{2}$$

where c_i is the concentration of i-th ion and z_i represents its charge. It is impossible to precisely calculate the ionic strength of the mitochondrial matrix. Taking into account just the inorganic ions present in the matrix (see Table 1) and assuming similar concentration of inorganic ions, one can assume I of the matrix to be in the range of 0.4 to 1 M. The extended Debye-Hückel can predict the activity coefficient in a solution with an ionic strength up to 100 mM and Davies equation is valid up to an ionic strength of 500 mM [42]. In this work, we use the modified Davies equation, which can correctly predict the values of the activity coefficient in solutions with ionic strength up to 1 M [42]:

$$\ln \gamma_i = -\frac{A z_i^2 \sqrt{I}}{1 + a_i B \sqrt{I}} + \frac{(0.2 - 4.17 \times 10^{-5} I) A z_i^2 I}{\sqrt{1000}} \tag{3}$$

where $A = \frac{\sqrt{2}F^2 e_0}{8\pi(\varepsilon RT)^{3/2}}$ and $B = \sqrt{\frac{2F^2}{\varepsilon RT}}$, F is the Faraday constant (96,488.46 C/mol), R is the ideal gas constant (8.3143 J/mol/K), T is the temperature (310 K), e_0 stands for the electronic charge (1.602×10^{-19} C), $\varepsilon = \varepsilon_0 \varepsilon_r$ is the permittivity of the medium ($\varepsilon_0 = 8.854 \times 10^{-12}$ F/m, $\varepsilon_r = 80$ for water), and a_i represents the radius of hydrated ion (values according to Kielland [70]).

Table 1. Mitochondrial concentrations of selected inorganic ions.

	Value	Measuring Technique	Type of Cell	Ref.
pH	7.14	SNARF	pig heart	[32]
	7.8	BCECF	beef heart	[43]
	7.65–8.1	SNARF	beef heart	[33]
	8.0–8.9	DMO distribution	rat heart	[39]
	7.40–7.51	DMO distribution	beef heart	[44]
	7.74–7.95	ratiometric pericam	motor nerve of *Drosophila* fly larvae	[40]
	7.7	SNARF	canine kidney (MDCK)	[34]
	7.64	YFP (SypHer)	HeLa	[38]
	7.6–8.2	GFP (mtAlpHi)	HeLa	[36]
	7.8	enhanced YFP	human bladder carcinoma	[45]
	7.7–8.2	pH-sensitive GFP	(Jurkat) T lymphocyte and HEK-293	[37]
	7.98	enhanced YFP	HeLa	[46]
	7.91	enhanced YFP	rat cardiomyocytes	[46]
	7.2–7.7	isotope distribution ‡	rat liver	[47]
	7.2–8.4	[U-^{14}C]glutamate efflux	rat kidney	[48]
Na$^+$	50–113 mM	CoroNa Red	canine kidney (MDCK)	[49]
	5 mM	SBFI	pig heart	[50]
	5.1 mM	SBFI	rat ventricular myocytes	[51]
	5 mM	assumed physiol. value	---------	[52]
K$^+$	15 mM	PBFI	isolated rat hepatocytes	[53]
	10–20 mM	PBFI	rat heart	[54]
	20–60 mM †	Potentiometric ISE	rat liver	[55]
	150 mM	assumed physiol. value	---------	[52]
Mg^{2+}	0.35 mM	Mg^{2+} efflux null-point	rat hepatocytes	[56]
	0.9 mM	Mg^{2+} efflux null-point	beef heart	[57]
	2.4 mM	^{31}P-NMR spectroscopy	heterotrophic sycamore	[58]
	0.5 mM	Fura-2	beef heart	[59]
	0.8–1.5 mM	Indo-1 and Fura-2	rat heart	[60]
Ca^{2+}	0.22 μM	mito-TN-XXL	motor nerve of *Drosophila* fly larvae	[40]
	24 μM *	Rhod-FF/Rhod-5N	motor nerve of *Drosophila* fly larvae	[40]
	0.19–3.34 μM	YC2	HEK-293 and HeLa	[61]
	0.41 μM	aequorin luminescence	bovine adrenal glomerulosa cells	[62]
	183 nM	Indo-1	rat heart	[63]
	0.1–0.6 μM	Indo-1/AM	rat cardiac myocytes	[64]
	0.2–1.1 μM	Indo-1	rat heart	[65]
	0.17–0.92 μM	Fura-2/AM	rat heart	[66]
	0.2–1.8 μM	Fura-2	rat heart	[67]
	0.4–2.1 μM	Fura-2	rat heart myocytes	[68]
	1–5 μM	Fura-2/AM	rat brain	[29]
Cl-	4.2 mM	liquid chromatography	human liver	[69]
	0.9–22.2 mM	liquid chromatography	human liver	[69]

* axon firing; † results originally given in mmol/mg; ‡ distribution of [^{14}C]– or [^{3}H]acetate or [^{14}C]5,5-dimethyloxazolidine-2,4-dione. Abbreviations: AM—acetoxymethyl ester, BCECF—2′,7′-biscarboxyethyl-5(6)-carboxyfluorescein, DMO—C-5-5-dimethyl-2,4-oxazolidinedione, GFP—green fluorescent protein, HEK-293—human embryonic kidney cells, Fura-2—calcium indicator C$_{29}$H$_{22}$N$_3$O$_{14}$K$_5$, HeLa—human cervix-carcinoma cells, Indo-1—calcium indicator C$_{32}$H$_{31}$N$_3$O$_{12}$, MDCK—Madin-Darby Canine Kidney, PBFI—potassium-binding benzofuran isophthalate, SBFI—sodium-binding benzofuran isophthalate, SNARF—seminaphthorhodafluors fluorescence, YC2—Yellow Cameleon protein, YFP—yellow fluorescent protein.

As presented in Figure 1, in the ionic strengths $I > 1$ mM, the ideal solution assumption ($\gamma_i = 1$) is not valid. In strong electrolytes, $I > 200$ mM, the activity coefficients ($\gamma_i \neq 1$) remain almost constant. In further calculations, the activity coefficients obtained for $I = 0.7$ M will be used.

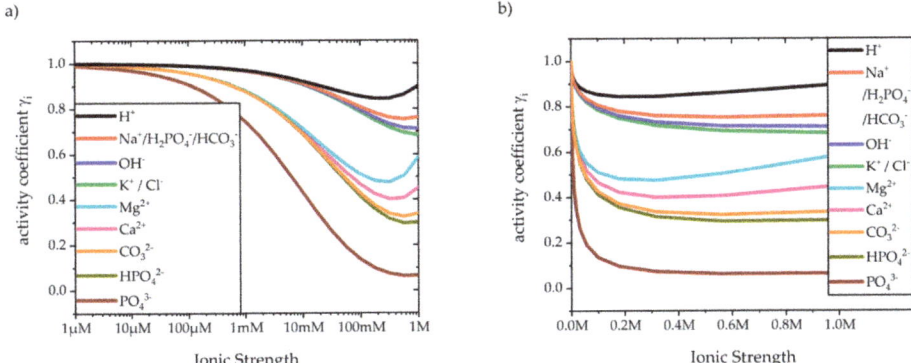

Figure 1. The calculated activity coefficients for the main inorganic ions present in mitochondria, (**a**) in logarithmic scale and (**b**) in linear scale.

2.2. Precipitation of Carbonates

Calcium and magnesium carbonates may form in the mitochondrial matrix due to the presence of carbon dioxide produced in citric acid cycle. Although the measurements of partial pressure of CO_2 inside mitochondria is not yet possible with existing methods, it might be assumed to be equal or slightly higher than the partial pressure in blood. The partial pressure of carbon dioxide measured in human mixed venous blood is $pCO_2 = 6.1$ kPa [71] to 6.3 kPa [72]. However, during exercises, it can rise to 7.8 kPa and after crossing the lactate acidosis threshold, it can reach values as high as 10.4 kPa [72].

The amount of gas dissolved in a solution is proportional to its partial pressure in the gas phase [73]. The proportionality factor is called the Henry's law constant. Nevertheless, it should be kept in mind that its value still depends on certain parameters, e.g., temperature and the ionic strength of the solution. The Henry's law constant for carbon dioxide dissolved in water at 25 °C equals $K_H = 3.4 \times 10^{-4}$ mol·m^{-3}·Pa^{-1} [74]. It lowers with the increasing temperature reaching 2.5×10^{-4} mol·m^{-3}·Pa^{-1} at 37 °C [75].

The hydration of carbon dioxide to form carbonic acid is slow to attain equilibrium below pH = 8 in pure systems. In biological systems the hydration of carbon dioxide is catalysed by carbonic anhydrase (a Zn-containing enzyme). Only a small portion of the aqueous carbon dioxide exists in its hydrated form as H_2CO_3. Hence, the relation between the carbonic species is usually expressed as:

$$\begin{cases} K_1 \times \{CO_2(aq)\} = \{H^+\} \times \{HCO_3^-\} \\ K_2 \times \{HCO_3^-\} = \{H^+\} \times \{CO_3^{2-}\} \end{cases} \quad (4)$$

where the equilibrium constants in pure water solutions at 25 °C equal $K_1 = 4.5 \times 10^{-7}$, $K_2 = 4.7 \times 10^{-11}$ [76].

These constants are temperature-dependent. Several different empirical equations have been developed to describe the relation between the equilibrium constants and temperature, including ones determined by Harned and co-workers [77–79], Ryzhenko [80], Millero and co-workers [81–83], as well as Plummer and Busenberg [75]. The values calculated using the latter at 37 °C, i.e., $K_1 = 4.96 \times 10^{-7}$ and $K_2 = 5.73 \times 10^{-11}$ are used in this work.

It should be noted, that mentioned equations are valid only for the solutions of carbon dioxide in pure water. The presence of other ions in the system lead to the increase of K_1 and K_2, e.g., values of equilibrium constants measured in sea water at 40 °C can reach $K_1 = 2.1 \times 10^{-6}$ and $K_2 = 2.3 \times 10^{-10}$ [83], comparing with calculated $K_1 = 5.1 \times 10^{-7}$ and $K_2 = 6.0 \times 10^{-11}$. In biological systems they can vary significantly and are treated not as true thermodynamic constants, but rather as apparent dissociation constants determined experimentally from the measurements of pH, pCO_2 and dissolved CO_2 [84]. Several studies [85–89] revealed the variability of K_1 in human plasma in the range of 4.78×10^{-7} to 1.58×10^{-6}. A case study of a diabetic child in ketoacidosis showed a change of K_1 from 3.24×10^{-6} to 9.54×10^{-7} after just 7 h of treatment [90].

2.2.1. Calcium Carbonates

Ca^{2+} and CO_3^{2-} ions form calcium carbonates $CaCO_3$. There are six different forms of $CaCO_3$, namely:

(1) calcite, anhydrous $CaCO_3$ with trigonal structure
(2) aragonite, anhydrous $CaCO_3$ with orthorhombic structure
(3) vaterite, anhydrous $CaCO_3$ with hexagonal structure
(4) monohydrocalcite, $CaCO_3 \cdot H_2O$ with trigonal structure
(5) hexahydrate known as ikaite, $CaCO_3 \cdot 6H_2O$ with monoclinic structure
(6) amorphous calcium carbonate (ACC), colloidal hydrate containing less than one molecule of water per molecule of $CaCO_3$.

Although the first attempts to quantify the solubility of calcium carbonates in aqueous systems date back to the middle of the 19th century [91,92] and despite the many efforts with current advanced ionic interaction models available, there are still no universally accepted values of the solubility constants of the calcium carbonate forms [93]. Values of the solubility products for various forms of $CaCO_3$ and their activity products are presented in Table 2. Using these values, one can calculate the equilibrium concentration of free calcium in the solutions of respective salts as functions of pH and partial pressure of carbon dioxide.

Table 2. The precipitation of insoluble calcium carbonates. The activity products (IAP) and the solubility products K_{sp} (at 37 °C) are obtained from the literature.

	Formula	Activity Product (Q)	Solubility Product, K_{sp} at 37 °C	Ref.
Calcite	$CaCO_3$	$\{Ca^{2+}\}\{CO_3^{2-}\}$	3.31×10^{-9} * 3.27×10^{-9} to 4.02×10^{-9}	[75] [93]
Aragonite	$CaCO_3$	$\{Ca^{2+}\}\{CO_3^{2-}\}$	4.57×10^{-9} * 4.67×10^{-9} to 5.41×10^{-9}	[75] [93]
Varietite	$CaCO_3$	$\{Ca^{2+}\}\{CO_3^{2-}\}$	1.23×10^{-8} * 1.18×10^{-8} to 1.84×10^{-8}	[75] [93]
Ikaite	$CaCO_3 \cdot 6H_2O$	$\{Ca^{2+}\}\{CO_3^{2-}\}$	2.40×10^{-7} * 3.50×10^{-7}	[94] [95]
Monohydro-calcite	$CaCO_3 \cdot H_2O$	$\{Ca^{2+}\}\{CO_3^{2-}\}$	7.09×10^{-8} * 2.51×10^{-8}	[96] [97]
ACC	$CaCO_3 \cdot nH_2O$ (n < 1)	$\{Ca^{2+}\}\{CO_3^{2-}\}$	9.09×10^{-7} * 3.98×10^{-7}	[94] [98]

* Values used in the calculations presented in this work.

2.2.2. Magnesium Carbonates

Mg^{2+} and CO_3^{2-} ions can form a variety of anhydrous, hydrated and basic magnesium carbonates, namely [99–103]:

(1) magnesite, anhydrous $MgCO_3$ with trigonal crystal structure
(2) barringtonite, dihydrate $MgCO_3 \cdot 2H_2O$ with triclinic structure
(3) nesquehonite, trihydrate $MgCO_3 \cdot 3H_2O$ with monoclinic structure
(4) lansfordite, pentahydrate $MgCO_3 \cdot 5H_2O$ with monoclinic structure
(5) pokrovskite, basic carbonate $Mg_2CO_3(OH)_2$ with monoclinic structure
(6) artinite, hydrated basic carbonate $Mg_2CO_3(OH)_2 \cdot 3H_2O$ with monoclinic structure
(7) hydromagnesite, hydrated basic carbonate $Mg_5(CO_3)_4(OH)_2 \cdot 4H_2O$ with monoclinic structure
(8) dypingite, hydrated basic carbonate $Mg_5(CO_3)_4(OH)_2 \cdot 5H_2O$ with monoclinic structure
(9) giorgiosite, hydrated basic carbonate $Mg_5(CO_3)_4(OH)_2 \cdot 6H_2O$
(10) octahydrate (UM1973-06-CO:MgH) with the formula $Mg_5(CO_3)_4(OH)_2 \cdot 8H_2O$, possibly identical to dypingite, but differs in optical properties
(11) protomagnesite, hydrated basic carbonate $Mg_5(CO_3)_4(OH)_2 \cdot 11H_2O$
(12) shelkovite, hydrated basic carbonate $Mg_7(CO_3)_5(OH)_4 \cdot 24H_2O$

Not all of the salts listed above can be precipitated directly from water solutions. Barringtonite is formed as the result of cold meteoric water percolating through olivine basalt and leaching magnesium from it [104]. Artinite is found along with hydromagnesite and other minerals as a low-temperature alteration product, as veinlets or crusts in serpentinized ultrabasic rocks [103]. Pokrovskite occurs within ultramafic bodies of dunite or serpentinite [105], and can be formed as an intermediate product in the thermal decomposition of artinite [106]. Shelkovite is a substance of anthropogenic origin, i.e., a product of burning coal mine dump [103].

Formation of the most stable anhydrous magnesite ($MgCO_3$) is strongly kinetically inhibited because of the high hydration energy of Mg^{2+} [107,108]. The precipitation of magnesite is apparently inhibited at temperatures of less than 80 °C [109–111]. Therefore, various hydrous magnesium carbonates form instead. In water solutions, landsfordite, nesquehonite and hydromagnesite can precipitate, depending on the temperature. Lansfordite decomposes into nesquehonite at temperatures above 10 °C [112,113]. Most commonly, only the mineral nesquehonite can be precipitated from aqueous solutions at 25 °C and partial pressure of CO_2 close to the ambient pressure or below [100,112–116]. At higher temperatures, i.e., above approximately 40 °C, various basic Mg-carbonates are usually formed by precipitation, mostly in the form of hydromagnesite [112,114,116,117].

The transformation from nesquehonite to hydromagnesite is unlikely to follow a single simple sequential reaction pathway [107] and a range of intermediate phases may form during this transformation [99]. Giorgiosite is a very poorly described basic magnesium carbonate [103,125] and it is considered as a metastable phase between nesquehonite and hydromagnesite [103]. Protohydromagnesite, a phase similar to dypingite, appears only as a metastable intermediate phase [114,126]. One intermediate phase is of particular interest, namely dypingite, a mineral that is known to be produced by cyanobacteria from alkaline wetlands [127]. Octahydrate $Mg_5(CO_3)_4(OH)_2 \cdot 8H_2O$ is often considered a form of dypingite [120].

The reactions leading to the formation of the relevant magnesium carbonates together with the values of respective solubility products at 25 °C are summarized in Table 3. Several values obtained at 35 °C are also included in Table 3. Using these values, one can calculate the equilibrium concentration of free magnesium in the solutions of respective salts as functions of pH and partial pressure of carbon dioxide.

Table 3. The precipitation of insoluble magnesium carbonates. The activity products (Q) and the solubility products K_{sp} (at 25 °C) are obtained from the literature.

	Formula	Activity Product (Q)	Solubility Product, K_{sp} at 25 °C	Ref.
Nesquehonite	$MgCO_3 \cdot 3H_2O$	$\{Mg^{2+}\}\{CO_3^{2-}\}$	2.38×10^{-6}	[99]
			1.10×10^{-5}	[118]
			2.57×10^{-6}	[119]
			5.37×10^{-6}	[120]
			4.57×10^{-6} *†	[120]
			5.90×10^{-5}	[121]
Artinite	$Mg_2CO_3(OH)_2 \cdot 3H_2O$	$\{Mg^{2+}\}^2 \{CO_3^{2-}\} \{OH^-\}^2$	6.31×10^{-18}	[119]
			7.76×10^{-18} *	[122]
Hydromagnesite	$Mg_5(CO_3)_4(OH)_2 \cdot 4H_2O$	$\{Mg^{2+}\}^5 \{CO_3^{2-}\}^4 \{OH^-\}^2$	6.31×10^{-31}	[119]
			1.25×10^{-32}	[123]
			3.16×10^{-33} *†	[123]
			8.32×10^{-38}	[124]
Dypingite	$Mg_5(CO_3)_4(OH)_2 \cdot nH_2O$ (n = 5 or 8)	$\{Mg^{2+}\}^5 \{CO_3^{2-}\}^4 \{OH^-\}^2$	1.12×10^{-35}	[120]
			9.12×10^{-37} *†	[120]

* Values used in the calculations presented in this work; † K_{sp} obtained at 35 °C.

2.3. Precipitation of Orthophosphates

Orthophosphate salts containing the PO_4^{3-} group are distinguished from metaphosphates and pyrophosphates that contain PO_3^- and $P_2O_7^{4-}$ groups, respectively. The PO_4^{3-} group takes part in ATP synthesis and occurs inside the mitochondrion in relatively high concentrations. Hence, the formation of metaphosphates and pyrophosphates will not be taken into consideration.

The concentration of phosphate ions is kept constant by the phosphate carrier while the relationship between them is given as:

$$\begin{cases} K_1 \times \{H_3PO_4\} = \{H^+\} \times \{H_2PO_4^-\} \\ K_2 \times \{H_2PO_4^-\} = \{H^+\} \times \{HPO_4^{2-}\} \\ K_3 \times \{HPO_4^{2-}\} = \{H^+\} \times \{PO_4^{3-}\} \end{cases} \quad (5)$$

where $K_1 = 7.41 \times 10^{-3}$, $K_2 = 6.31 \times 10^{-8}$ and $K_3 = 4.07 \times 10^{-13}$ [128].

2.3.1. Calcium Orthophosphates

The calcium ortophosphates found in nature include [30,129–131]:

(1) monocalcium phosphate monohydrate (MCPM), $Ca(H_2PO_4)_2 \cdot H_2O$ with Ca/P ratio = 0.5
(2) anhydrous monocalcium phosphate (AMCP), $Ca(H_2PO_4)_2$ with Ca/P ratio = 0.5
(3) dicalcium phosphate dihydrate, $CaHPO_4 \cdot H_2O$, known as brushite (BRU), Ca/P ratio = 1.0
(4) anhydrous dicalcium phosphate, $CaHPO_4$, known as monetite, Ca/P ratio = 1.0
(5) octacalcium phosphate (OCP) with the formula $Ca_8(HPO_4)_2(PO_4)_4 \cdot 5H_2O$, Ca/P ratio = 1.33
(6) α-tricalcium phosphate (α-TCP), $Ca_3(PO_4)_2$ with monoclinic crystallographic structure and Ca/P ratio = 1.5
(7) β-tricalcium phosphate (β-TCP), $Ca_3(PO_4)_2$ with rhombohedral crystallographic structure and Ca/P ratio = 1.5
(8) amorphous calcium phosphates (ACP), $Ca_xH_y(PO_4)_z \cdot nH_2O$, n = 3 to 4.5, Ca/P ratio = 1.2 to 2.2
(9) calcium-deficient hydroxyapatite (CDHA), $Ca_{10-x}(HPO_4)_x(PO_4)_{6-x}(OH)_{2-x}$ ($0 < x \le 2$) with Ca/P ratio = 1.5 to 1.67
(10) hydroxyapatite (HA) with the formula $Ca_{10}(PO_4)_6(OH)_2$, Ca/P ratio = 1.67
(11) chloroapatite (ClA) with the formula $Ca_{10}(PO_4)_6(Cl)_2$, Ca/P ratio = 1.67
(12) carbonated apatite (CO_3A), $Ca_{10}(PO_4)_6CO_3$, known as dahllite, Ca/P ratio = 1.67

(13) fluoroapatite (FA), with the formula $Ca_{10}(PO_4)_6F_2$ and Ca/P ratio = 1.67
(14) oxyapatite (OA), with the formula $Ca_{10}(PO_4)_6O$ and Ca/P ratio = 1.67
(15) tetracalcium phosphate (TetCP), $Ca_4(PO_4)_2O$, known as hilgenstockite, Ca/P ratio = 2.0

Another calcium salt, i.e., isoclasite with the formula $Ca_2(PO_4)OH\cdot 2H_2O$, has been reported in 1870 by Sandberger [132] as orthorhombic crystals. His measurement was rather inaccurate (the angle between prismatic faces measured by contact goniometer) and the crystals could be hexagonal, possibly of apatite. Doubts concerning validity of data on isoclasite are supported by the absence of any additional description of isoclasite anywhere [133].

Some of the calcium salts mentioned above, including tetracalcium phosphate, oxyapatite, α-tricalcium phosphate and β-tricalcium phosphate, cannot be precipitated from aqueous solutions [131]. Anhydrous monocalcium phosphate and anhydrous dicalcium phosphate are stable only in temperatures above 100 °C, while monocalcium phosphate monohydrate is stable only at pH < 2 [131]. Energy dispersive x-ray spectra from mitochondrial precipitates do not show the presence of fluor [27]. Hence, the formation of fluoroapatite can be ruled out as well.

The formulas of the calcium phosphates that might precipitate in the mitochondrion matrix, together with the values of respective solubility products at 37 °C are summarized in Table 4 (some of the literature values have been recalculated to correspond to the activity product given in the table). To calculate the equilibrium concentration of chloroapatite and carbonated apatite, the concentrations of chlorides and carbonates have to be provided. The free chloride concentration in the mitochondrial matrix equals $[Cl^-]$ = 4.2 mM [69]. The concentration of carbonate ions $[CO_3^{2-}]$ are calculated using Equation (4), assuming that partial pressure of carbon dioxide equals pCO_2 = 7 kPa.

Table 4. The precipitation of insoluble calcium orthophosphates. The activity products (Q) and the solubility products K_{sp} (at 37 °C) are obtained from the literature.

	Formula	Activity Product (Q)	Solubility Product, K_{sp} at 37 °C	Ref.
BRU	$CaHPO_4$	$\{Ca^{2+}\}\{HPO_4^{2-}\}$	2.34×10^{-7} * 2.26×10^{-7} 0.92×10^{-7}	[129] [134] [135]
OCP	$Ca_8(HPO_4)_2(PO_4)_4$	$\{Ca^{2+}\}^8\{H^+\}^2\{PO_4^{3-}\}^6$	1.25×10^{-96} * 1.20×10^{-97} 3.98×10^{-98} 2.51×10^{-99}	[129] [134] [136] [137]
ACP	$Ca_xH_y(PO_4)_z$	$\{Ca^{2+}\}^x\{H^+\}^y\{PO_4^{3-}\}^z$ $\{Ca^{2+}\}^3\{PO_4^{3-}\}^2$ $\{Ca^{2+}\}\{H^+\}^{0.22}\{PO_4^{3-}\}^{0.74}$	2.0×10^{-26} *† 2.0×10^{-33}–1.6×10^{-25} $2.3 \times 10^{-11}, 3.2 \times 10^{-12}$ §	[130] [29] [138]
CDHA	$Ca_{10-x}(HPO_4)_x(PO_4)_{6-x}(OH)_{2-x}$	$\{Ca^{2+}\}^{10-x}\{H^+\}^x\{PO_4^{3-}\}^6\{OH^-\}^{2-x}$	$\sim 7.94 \times 10^{-86}$ *	[130]
HA	$Ca_{10}(PO_4)_6(OH)_2$	$\{Ca^{2+}\}^{10}\{PO_4^{3-}\}^6\{OH^-\}^2$	1.00×10^{-118} 6.31×10^{-118} * 2.00×10^{-119} ‡ 2.51×10^{-111} 5.42×10^{-119}	[29] [129] [139] [140] [141]
ClA	$Ca_{10}(PO_4)_6Cl_2$	$\{Ca^{2+}\}^{10}\{PO_4^{3-}\}^6\{Cl^-\}^2$	3.98×10^{-116} *	[140]
CO_3A	$Ca_{10}(PO_4)_6CO_3$	$\{Ca^{2+}\}^{10}\{PO_4^{3-}\}^6\{CO_3^{2-}\}$	1.58×10^{-103} *‡	[139]

* Values used in the calculations presented in this work; † K_{sp} is pH-dependent, value in the table for pH = 7.4; ‡ K_{sp} obtained at 25 °C; § authors observed two different metastable phases.

2.3.2. Magnesium Orthophosphates

The magnesium ortophosphates found in nature include [142]:

(1) anhydrous trimagnesium phosphate, $Mg_3(PO_4)_2$ with monoclinic crystal structure, occurs as the natural mineral farringtonite
(2) trimagnesium phosphate octahydrate, $Mg_3(PO_4)_2\cdot 8H_2O$ with monoclinic structure, occurs as the natural mineral babierrite

(3) trimagnesium phosphate docosahydrate, $Mg_3(PO_4)_2 \cdot 22H_2O$ with triclinic structure, occurs as the natural mineral cattiite
(4) anhydrous dimagnesium phosphate, $Mg(HPO_4)$
(5) dimagnesium phosphate monohydrate, $Mg(HPO_4) \cdot H_2O$
(6) dimagnesium phosphate trihydrate, $Mg(HPO_4) \cdot 3H_2O$ with orthorhombic structure, occurs as the natural mineral newberyite
(7) dimagnesium phosphate heptahydrate, $Mg(HPO_4) \cdot 7H_2O$ with monoclinic structure, occurs as the natural mineral phosphorrösslerite
(8) anhydrous monomagnesium phosphate, $Mg(H_2PO_4)_2$
(9) monomagnesium phosphate dihydrate, $Mg(H_2PO_4)_2 \cdot 2H_2O$
(10) monomagnesium phosphate tetrahydrate, $Mg(H_2PO_4)_2 \cdot 4H_2O$
(11) magnesium hydroxyphosphate, $Mg_2(PO_4)(OH)$ with trigonal structure, occurs as the natural mineral holtedahlite
(12) magnesium ammonium phosphate, $NH_4MgPO_4 \cdot 6H_2O$ with orthorhombic structure, occurs as the natural mineral struvite
(13) magnesium potassium phosphate, $KMgPO_4 \cdot 6H_2O$

Due to the high hydration energy, the direct precipitation of anhydrous magnesium phosphates is not possible. The anhydrous $Mg(HPO_4)$ compound is formed by dehydration at 100–170 °C [142]. The anhydrous farringtonite is formed by heating its hydrates at 400 °C [142].

Monomagnesium phosphate and its hydrates dissolve in water, forming phosphoric acid and depositing a solid precipitate of $Mg(HPO_4) \cdot 3H_2O$ [142], e.g., precipitation monomagnesium phosphate tetrahydrate requires a non-aqueous solvent in which H_3PO_4 is soluble but $Mg(H_2PO_4)_2 \cdot 4H_2O$ has limited solubility [143]. Among dimagnesium phosphates only the trihydrate is stable in the $MgO/H_2O/P_2O_5$ system at room temperature [142]. The trimagnesium phosphate docosahydrate is metastable and reverts to octahydrate in water [144].

It has been shown that increased levels of ammonia disturb mitochondrial function, induce oxidative stress, causes mitochondrial membrane potential collapse and negatively affects several key enzymes responsible for energy metabolism [145–148]. Although the mitochondria are responsible for the production of NH_4^+ ions during the oxidation of glutamate [149], in healthy mitochondria the concentration of ammonia is very low. Hence, the formation of struvite can be ruled out. Furthermore, in the biological precipitation of struvite, Ca^{2+} ions make struvite precipitation difficult and even inhibit it [150,151].

The reactions leading to the formation of the relevant magnesium phosphates together with the values of respective solubility products at 38 °C are summarized in Table 5. Using these values, the equilibrium concentration of phosphates $[\sum P_i]_{eq}$, required for the formation of each magnesium orthophosphate can be calculated.

Table 5. The precipitation of insoluble magnesium phosphates. The activity products (Q) and the solubility products K_{sp} (at 38 °C) are obtained from the literature.

	Formula	Activity Product (Q)	Solubility Product, K_{sp} at 38 °C	Ref.
Babierrite	$Mg_3(PO_4)_2 \cdot 8H_2O$	$\{Mg^{2+}\}^3 \{PO_4^{3-}\}^2$	6.31×10^{-28}	[152]
			5.37×10^{-26} *†‡	[144]
			2.00×10^{-26} §	[153]
Newberyite	$Mg(HPO_4) \cdot 3H_2O$	$\{Mg^{2+}\} \{HPO_4^{2-}\}$	3.47×10^{-5}	[154]
			1.78×10^{-6} *†	[144]
			2.40×10^{-6} ‡	[144]
			1.78×10^{-6} §	[155]
			1.48×10^{-6} §	[153]
Magnesium potassium phosphate	$KMgPO_4 \cdot 6H_2O$	$\{K^+\} \{Mg^{2+}\} \{PO_4^{3-}\}$	2.4×10^{-11} §	[156]

* Values used in the calculations presented in this work; † recalculated based on [152]; ‡ recalculated based on [154]; § K_{sp} obtained at 25 °C.

2.4. Precipitation of Polyphosphates

It has been reported that PO_4^{3-} can be concentrated and stored by biochemical polymerization into polyphosphate (polyP) ions [157]. They are negatively charged polyanions $(P_nO_{3n+1})^{(n+2)-}$ with great affinity for calcium and other multivalent cations [158]. PolyP is a macroergic compound so that the energy of phosphoanhydride bond hydrolysis is the same as in ATP [159]. Polyphosphates have been found in all organisms ranging from bacteria to mammals [160].

The synthesis of polyP is mainly realized by the polyphosphate kinase, which catalyses the reaction $ATP + [PO_3^-]_n \leftrightarrow ADP + [PO_3^-]_{n+1}$. PolyP can play a role in the energetic metabolism of mitochondria [161], e.g., it has been shown that, in human cell lines, specific reduction of mitochondrial polyP under expression of yeast exopolyphosphatase PPX1 significantly modulates mitochondrial bioenergetics and transport [159]. Polyphosphates (including pyrophosphate) have also a significant role in the regulation of the intensity of several biosynthetic processes in mitochondrial and cell energy metabolism [162,163].

PolyP can form several insoluble salts. Bobtelsky and Kertes [164], have analysed the properties of pyrophosphates $(P_2O_7)^{4-}$ and tripolyphosphates $(P_3O_{10})^{5-}$ and the precipitation of calcium, magnesium, barium and strontium salts of these ions. With the addition of calcium to the respective polyP solution, the insoluble $Ca_2P_2O_7$ and $Ca_5(P_3O_{10})_2$ will form, while the addition of magnesium leads to the precipitation of $Na_2Mg_3(P_2O_7)_2$ and $NaMg_2P_3O_{10}$, respectively. Interestingly, the redissolution of these salts will take place if the excess of the polyphosphate ions is added to the solution [164].

Polyphosphate chelation of calcium reduces the free calcium concentration [165] and produces a neutral, amorphous, $[Ca(PO_3)_2]_n$ complex. Discrete granules containing calcium–polyP complexes have been observed e.g., in acidocalcisomes, the electron-dense acidic organelles that have been conserved from bacteria to humans [166]. PolyP are noted to be highly inhibitory to calcium phosphate nucleation and precipitation [167], e.g., they inhibit crystalline calcium hydroxyapatite nucleation and growth from solution [168]. However, in basic pH solutions calcium polyphosphate glass releases hydrated calcium and polyphosphate ions into solution and, through hydrolytic degradation of polyP, calcium polyphosphate eventually transforms into hydroxyapatite [169]. The relationships between polyphosphate chemistry and apatite biomineralization as well as the possible role of polyP in mitochondria have been discussed in detail by Omelon et al. [165,170].

The important polyP forms in living cells are specific complexes with polyhydroxybutyrate (PHB) and Ca^{2+}, which have been found in the membranes of many organisms, including membranes of mitochondria of animal cells [171,172]. The PHB forms a helix with a lipophilic shell of methyl groups and polar lining of ester carbonyl oxygens surrounding a core helix of polyP with Ca^{2+} bridging the two polymers. These helices can form voltage-dependent, Ca^{2+}-selective channels, or be components of ion-conducting proteins: namely, the human erythrocyte Ca^{2+}-ATPase pump and the *Streptomyces lividans* potassium channel [173].

The polyP pool in mitochondria likely consists of two parts, i.e., one part localized directly in the membranes as a polyP–PHB–Ca^{2+} complex, and the other part present in the intermembrane space and not bound to PHB, though its association with cations, including Ca^{2+}, is not improbable [159]. It was determined by subfractionation of isolated mitochondria by mild osmotic treatment, that intermembrane space contained ~90% of total mitochondrial polyP [174]. This suggests that polyphosphates are formed outside the mitochondrial matrix. Furthermore, the formation of calcium polyP inside the matrix seems less likely than calcium orthophosphates, because the Ca/P ratio of this complex is less than 1, while the mitochondrial Ca/P ratio is between 1 and 2.

3. Results

3.1. Precipitation of Calcium Carbonates

Using the values presented in Table 2, one can calculate the equilibrium concentration of free calcium in the solutions of respective salts as functions of pH and partial pressure of carbon dioxide. Such calculations are shown in Figure 2.

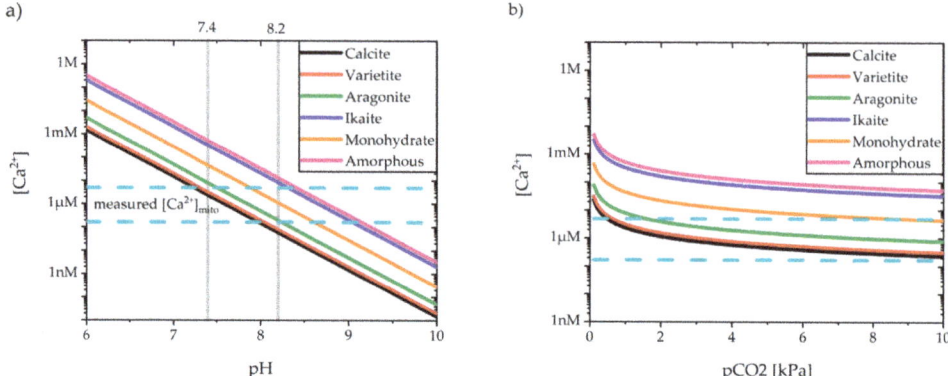

Figure 2. The calculated equilibrium concentrations of free calcium (**a**) as functions of pH at 37 °C and pCO_2 = 7 kPa, (**b**) as a function of partial pressure of carbon dioxide at 37 °C and pH = 7.8.

The more stable crystalline forms are generally those with a lower solubility product, hence lower equilibrium concentration of calcium. However, the formation and transformation of solid phases during the spontaneous precipitation of calcium carbonate from alkaline, carbonate-rich solutions is a highly complicated and time-dependent process. The local conditions of temperature, concentration and impurities can have a profound effect on the qualitative nature of the phases formed, as well as on the quantitative kinetics of individual processes.

The Ostwald-Lussac Law of Stages states that, in the pathway to the final crystalline state, will pass through all less stable states in order of increasing stability [175]. The formation of metastable phases is a quite common phenomenon during spontaneous precipitation of supersaturated solutions [98,176,177]. During the precipitation of more stable anhydrous forms, amorphous calcium carbonate, calcium carbonate hexahydrate and calcium monohydrocalcite appear as intermediate phases of importance and consequence. At high supersaturation, the first-formed phase is ACC, a rather unstable precursor consisting of spherical particles of calcium carbonate [94,98]. ACC changes to more stable forms via dissolution and precipitation. The solubility product of ACC is higher than for other forms of $CaCO_3$. Hence, any solution in equilibrium with the amorphous phase will be highly supersaturated with respect to the crystalline phases, which leads to nucleation-controlled growth [178]. In other words, ACC serves as a precursor for other polymorphs, the nucleation of the polymorphous crystals occurs in the vicinity of the clouded precursor and the crystals grow by consuming the precursors surrounding the crystals [179].

If magnesium ions are present in the reaction system, they become incorporated into the ACC structure. The crystalline phase, which forms from the amorphous precursor, is controlled by the magnesium content of the precursor. Pure crystallises to vaterite [180], ACC containing 10% Mg crystallises to calcite [181], ACC with 30% crystallises to monohydrocalcite [178,182] and ACC with 50% Mg crystallises to protodolomite/dolomite (at temperatures higher than 60 °C) [183].

High concentrations of orthophosphates prevent the crystallization of the more stable anhydrous forms of $CaCO_3$, which leads to the occurrence of ikaite and monohydrocalcite in the system [94,98]. Ikaite, however, for its formation requires near-freezing conditions and is unstable in the temperatures above 25 °C [184]. Soluble acidic proteins occurring in the mitochondrial matrix might also have a strong inhibiting effect on the nucleation of anhydrous polymorphs, e.g., it has been shown that poly-L-aspartic acid completely inhibits the formation of varietite [185]. Na^+ and Cl^- ions exert a weak or no influence on the precipitation of carbonates [186,187].

The presence of both magnesium and orthophosphate ions in mitochondrial matrix suggests that monohydrocalcite might be the phase occurring inside mitochondria. However, the solubility calculations presented in Figure 2, clearly show that monohydrocalcite can form only if matrix pH is greater than 7.8 (for [Ca^{2+}] = 5 µM) or even 8.6 (for [Ca^{2+}] = 0.17 µM). In lower pH, neither monohydrocalcite, nor any other calcium carbonate will precipitate.

3.2. Precipitation of Magnesium Carbonates

Using the solubility products presented in Table 3, one can calculate the equilibrium concentration of free magnesium in the solutions of respective salts as functions of pH and partial pressure of carbon dioxide. Solubility calculations, presented in Figure 3, clearly show that nesquehonite is the most stable magnesium carbonate in given conditions. It is also the only magnesium carbonate that can precipitate in the mitochondrial matrix, but only if the matrix pH is greater than 7.5 (for [Mg^{2+}] = 1.5 mM) or even 7.9 (for [Mg^{2+}] = 0.35 mM). In lower pH, neither nesquehonite, nor any other magnesium carbonate will precipitate.

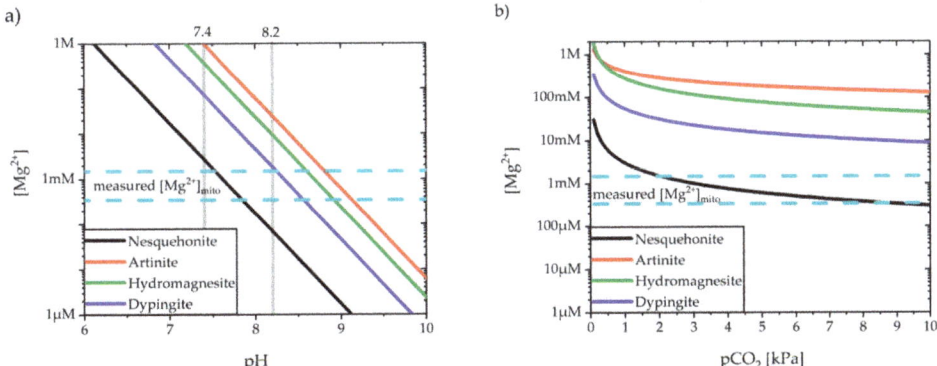

Figure 3. The calculated equilibrium concentrations of free magnesium (**a**) as functions of pH at 35 °C and pCO$_2$ = 7 kPa, (**b**) as a function of partial pressure of carbon dioxide at 35 °C and pH = 7.8.

3.3. Precipitation of Calcium Orthophospates

The total concentration of phosphates [ΣP$_i$]$_{eq}$, required for the formation of each salt at constant free calcium concentration [Ca^{2+}] = 0.17 µM and [Ca^{2+}] = 5 µM, is presented in Figure 4. Salts with the lowest [ΣP$_i$]$_{eq}$ are thermodynamically most stable and therefore dominate in the system, i.e., whenever the total concentration of free phosphates exceeds the lowest [ΣP$_i$]$_{eq}$, formation of the respective salt will occur, keeping [ΣP$_i$] at a constant level. As shown in Figure 4, carbonated apatite is the most stable calcium salt, if [Ca^{2+}] is kept in the micromolar range.

Assuming the constant concentration of phosphates in the cytoplasm [ΣP$_i$]$_{cyt}$ = 2 mM, and that the equilibrium at phosphate translocase yields [ΣP$_i$]$_{mat}$/[ΣP$_i$]$_{cyt}$ = [H$^+$]$_{cyt}$/[H$^+$]$_{mat}$, one can calculate the total concentration of mitochondrial phosphates [30]. The calculated values of [ΣP$_i$]$_{mat}$ compared with the [ΣP$_i$]$_{eq}$ of the respective salts (see Figure 4b) clearly show that if the calcium concentration is elevated, the solution of the mitochondrion matrix (pH = 7.4 to 8.2) is supersaturated with respect to up to three salts. i.e., hydroxyapatite, chloroapatite and carbonated apatite. Other salts (OCP, brushite, CDHA and ACP) having [ΣP$_i$]$_{eq}$ > [ΣP$_i$]$_{mat}$ will dissolve, hence cannot form inside mitochondria.

Calcium apatites (hydroxyapatite, chloroapatite, fluoroapatite and carbonated apatite) are members of a large family of more than 75 chemically different apatites, i.e., compounds with the chemical formula: A$_2$B$_3$(XO$_4$)$_3$Y, where A and B are bivalent cations, and XO$_4$ and Y are trivalent and monovalent anions, respectively [188]. The high-symmetry members of the apatite family crystallise in the hexagonal system.

Calcium apatites are very often non-stoichiometric, due to the substitution of OH$^-$ in hydroxyapatite with Cl$^-$, F$^-$ and CO$_3^{2-}$. Therefore, the formulas for carbonated apatite and chloroapatite are often given as Ca$_{10}$(PO$_4$)$_6$(CO$_3$)$_x$(OH)$_{2-2x}$ and Ca$_{10}$(PO$_4$)$_6$Cl$_x$(OH)$_{2-x}$, respectively [139,140].

The apatite structure allows varied substitutions of calcium and phosphate ions to take place without a significant alteration in its basic structure, e.g., one of the basic phosphorite minerals, known as francolite, has a variable chemical composition, which can be represented by $(Ca,Mg,Sr,Na)_{10}(PO_4,SO_4,CO_3)_6F_{2+x}$, where $0 < x < 1$ [189,190].

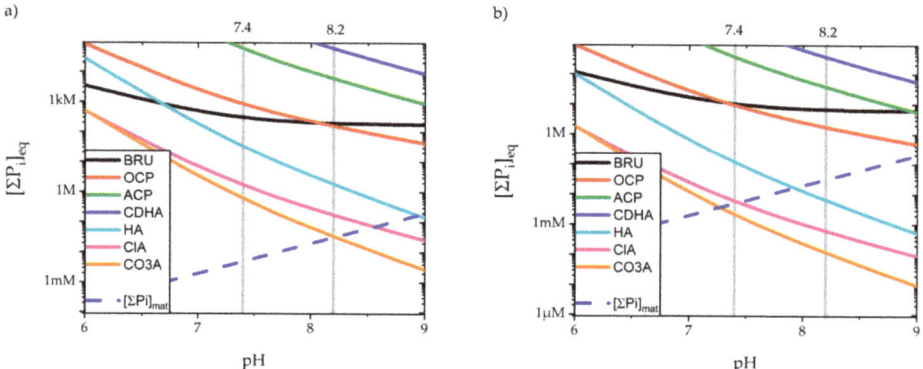

Figure 4. The calculated equilibrium concentrations $[\Sigma P_i]_{eq}$ required for the formation of calcium phosphates as functions of pH. Calcium concentration set to (**a**) $[Ca^{2+}] = 0.17\ \mu M$ and (**b**) $[Ca^{2+}] = 5\ \mu M$.

Formation of CO_3^{2-}-substituted hydroxyapatite is common in biological and geological systems and this substitution has been well documented in several studies [191–193]. Carbonate ion in the structure of biological apatites can substitute either the OH^- ion (type A carbonated apatite) or PO_4^{3-} ion (type B carbonated apatite) [194] and these two locations of carbonate species can be distinguished using spectroscopic methods [195,196]. The carbonate group distorts the apatite lattice and causes the resulting solid to be considerably more soluble than hydroxyapatite [197]. The substitution of carbonate for phosphate would also reduce the crystallinity and precipitation rate of HA [198]. Another specificity of biological apatites is the presence of hydrogen phosphate (HPO_4^{2-}) ions in PO_4^{3-} sites [199–201]. These two types of bivalent ions substituting for PO_4^{3-} have been shown to correspond to the formation of calcium deficient apatites, i.e., the loss of a negative charge due to these substitutions is compensated by the creation of cationic vacancies and anionic vacancies in the OH^- site [202]. The formula for biological apatites taking into account the possible existence of both type A and B substitutions reads [203,204]:

$$Ca_{10-x}(PO_4)_{6-x}(HPO_4 \text{ or } CO_3)_x(OH \text{ or } \frac{1}{2}CO_3)_{2-x} \text{ with } 0 < x < 2. \qquad (6)$$

This general formula can be used, e.g., to represent the bone composition at all ages in many vertebrates ($x = 1.7$) or a typical composition of human tooth enamel ($x = 0.6$) [203]. The Ca/P ratio of carbonated apatite vary from 2.0 ($x = 2$ and all type B substitutions with CO_3^{2-}) down to 1.2 ($x = 2$ and all type B substitutions with HPO_4^{2-}) or even lower if cationic substitutions ($2Na^+$ or Mg^{2+} for Ca^{2+}) occur.

Precipitated apatite crystals generally exhibit a structured hydrated surface layer containing mainly bivalent cations and anions. This feature should not be confused with the Stern double layer that forms on most mineral surfaces. The substitution possibilities in the hydrated layer are not well known but seem to be greater than in the core of the apatite structure [203]. Magnesium ions for example, which hardly penetrate the apatite lattice, are easily and reversibly incorporated into the hydrated layer in large amounts [205]. Ion substitutions in the hydrated layer considerably modify its structure even though these alterations seem reversible in most cases [206].

3.4. Precipitation of Magnesium Orthophospates

The solubility calculations for magnesium orthophosphates at low magnesium concentration $[Mg^{2+}] = 0.35$ mM (see Figure 5a) clearly show that solution of mitochondrial matrix is undersaturated with the respect to all magnesium phosphate salts. Hence, they cannot precipitate. At high magnesium concentrations $[Mg^{2+}] = 1.5$ mM, the precipitation of babierrite is possible if pH > 7.9 (see Figure 5b). In lower pH, neither babierrite, nor any other magnesium phosphate will precipitate.

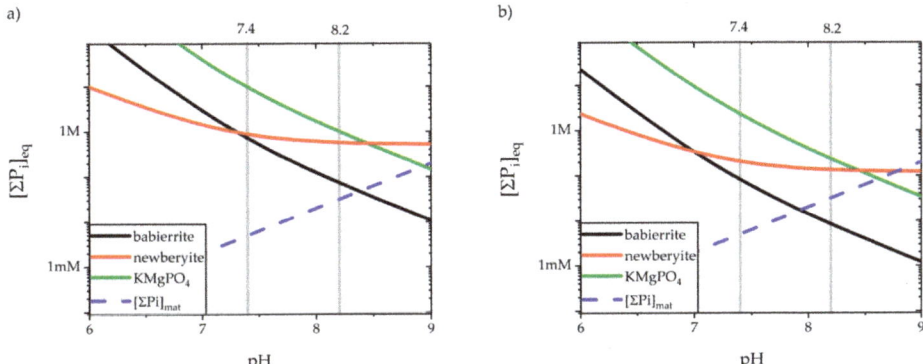

Figure 5. The calculated equilibrium concentrations $[\sum P_i]_{eq}$ required for the formation of magnesium phosphates as functions of pH. Magnesium concentration set to (**a**) $[Mg^{2+}] = 0.35$ mM and (**b**) $[Mg^{2+}] = 1.5$ mM.

4. Discussion

The results of solubility calculations (presented in Figures 2–5) show that four different salts can form in the mitochondrion, depending on the changes in ionic concentrations induced by the metabolic processes in mitochondria and by the concentration changes in cytosol. Babierrite can precipitate at high magnesium concentration and pH > 9. Nesquehonite precipitates at pH > 7.5 (or pH > 7.8 if calcium concentration is low). Monohydrocalcite precipitates at high Ca^{2+} concentrations and pH > 7.8. Carbonated apatite at high calcium concentrations precipitates in a whole mitochondrial pH range, and dissolves when mitochondrial $[Ca^{2+}]$ is low. However, these calculations do not take into account the influence of magnesium ions on precipitation of calcium salts and vice versa. They also neglect the influence of phosphate ions on the precipitation of carbonate salts and the influence of carbonate ions on the precipitation of phosphate salts. These relations are discussed below.

4.1. Carbonates in the Ca/Mg System

Insoluble salts containing both calcium and magnesium are known as dolomites, i.e., the family of non-stoichiometric compounds with the formula $Ca_{1-x}Mg_{1+x}(CO_3)_2$ [207–210]. Sedimentary dolomites with x < 0 are named calcium dolomites, and the ones with x > 0 are named magnesian dolomites. The carbonate dominated by calcite is called limestone and the mineral $Ca_{0.5}Mg_{1.5}(CO_3)_2$ (x = 0.5) is named huntite. Other carbonates: ankerite $Ca(Fe, Mg, Mn)(CO_3)_2$ and kutnohorite $Ca(Mn,Mg,Fe)(CO_3)_2$, with bivalent cations: Fe^{2+} and Mn^{2+}, substituting a part of Mg^{2+} ions in the layered structure, also enter the dolomite group [211].

The hydration shell of magnesium is more strongly bound than the hydration shell of calcium [108,212]. Hence, the energy required to desolvate magnesium is higher than that of calcium. That explains the high temperatures required to crystallize anhydrous Mg-Ca carbonates, i.e., temperature of 60 °C is required for the formation of dolomite [183] and repeated trials to precipitate it under laboratory conditions at room temperature were unsuccessful [213,214]. Another anhydrous carbonate with two cations, namely eitelite with the formula $Na_2Mg(CO_3)_2$, can also be precipitated only at temperatures above 60 °C [215].

The hydrated nature of monohydrocalcite means that full dehydration of Mg^{2+} is not required before incorporation into the crystal lattice. Therefore, it will more likely form than the anhydrous calcium carbonate phases [178]. Monohydrocalcite has a high capacity to accommodate magnesium within its structure. The coordination number of Ca^{2+} in monohydrocalcite is 8 [216,217] and Mg^{2+} does not take a coordination number other than 6. Therefore, Mg^{2+} is not incorporated into the monohydrocalcite structure, directly but forms discrete hydrated magnesium carbonate insertions. According to the solubility calculations presented in Figures 2 and 3, the solution in mitochondrial matrix is saturated with respect to both monohydrocalcite and nesquehonite, if pH > 8. It is in agreement with the observations of Nishiyama et al. [182], who have shown that in $Ca^{2+}/Mg^{2+}/CO_3^{2-}$ systems nesquehonite with very low crystallinity is expected to be formed with monohydrocalcite and the amorphous material. The hydrous magnesium carbonates surrounding monohydrocalcite play a protective role in preventing its dehydration to anhydrous calcium carbonates.

High-Mg monohydrocalcite consists of individual nanometer-sized crystals (<35 nm) with a significant part of its structural H_2O being associated with magnesium [178]. X-ray diffraction (XRD) experiments show that monohydrocalcite exhibits low crystallinity and small particle size when the ratio of magnesium to calcium in the solid is higher than 0.4, i.e., the increase of magnesium content lead to broad X-ray diffraction peaks with low intensity. Monohydrocalcite with Mg/Ca ratio higher than 0.5 exhibits the properties characteristic for amorphous material (no XRD peaks are observed) [182]. This corresponds well with the results of the electron microscopy measurements of brain and liver mitochondria that show a lack of crystal structures in the matrix [27]. The value of solubility of amorphous Mg-rich monohydrocalcite is slightly higher than in the case of pure monohydrocalcite [182]. Hence, in order to precipitate, it requires the solution to be more alkaline than the mitochondrial matrix.

4.2. Carbonated Apatite in Physiological Solutions

The studies on slow spontaneous precipitation of calcium carbonates and calcium phosphates under physiological conditions [218], show that calcium carbonate precipitation is prevented by the presence of phosphate ions at concentrations too low for the precipitation of calcium phosphate. Such precipitation is prevented, when the phosphate/bicarbonate concentration ratio is higher than about 1/300. Above this ratio, the competition of the phosphate ions with the carbonate ions is sufficient to inhibit the nucleation and crystal growth of calcium carbonates, while the phosphate concentration is still too low for the nucleation of phosphates [218]. Pyrophosphate and hexametaphosphate ions inhibit the precipitation of calcium carbonate in a similar way. Hence, the composition of most biological fluids does not allow the deposition of calcium carbonate and it is suggested that such deposition can occur only under the influence of metabolic processes which locally raise the bicarbonate and lower the phosphate concentration [218].

Carbonate ions may affect the solubility of phosphate salts in several ways [191]. Ion pair complexes between calcium and bicarbonate and carbonate will form and the presence of ion pairs and reduction of free Ca^{2+} would increase the soluble concentration of phosphate. However, the calculations indicate that ion pairs will be the predominant calcium-containing species at pH greater than 8.5 and at high carbonate levels (0.05 M) [219]. The carbonate ions may also compete with phosphate for sites on the growing crystallites. Nucleation and crystal growth may be hindered by the extra time and energy required to eliminate carbonate from phosphate sites and vice versa. When crystal growth does take place, carbonate can readily substitute for phosphate in the apatite lattice, as discussed earlier. Chemical analyses reveal that the apatitic tissues always contain smaller or larger amounts of carbonate which does not crystallize separately as carbonate [197].

Precipitation of babierrite is also inhibited, because of the presence of carbonate and calcium ions. Due to the chemical similarity between Mg^{2+} and Ca^{2+}, magnesium co-precipitates with calcium phosphates instead, inducing the formation of an amorphous phase rather than a stable crystalline carbonated apatite [191,220]. Magnesium ions stabilize the amorphous precipitates of calcium carbonate phosphate. Apatitic precipitates are more poorly-crystallized when formed in the presence of magnesium ions than when formed in their absence, and the crystallinity of the precipitates formed in the presence of magnesium decreases with rising [Mg^{2+}]. The Ca/P ratios of the amorphous precipitates are often significantly lower when formed in the presence of Mg^{2+} ions, and the equilibrium concentrations of calcium and phosphates are higher when the precipitates are amorphous, than when they are crystalline [221].

Biological calcium apatites show a broad range of chemical compositions but also structure. It has been shown that hydroxyapatite precipitates obtained in the conditions where pH is kept constant show low crystallinity [222]. In primitive organisms, calcium apatites are generally amorphous, while in more elaborated organisms, especially in vertebrates, calcium apatites can crystallise [203]. Magnesium has been shown to disturb the crystallization of calcium apatites from solution when used at concentrations sufficient to be a major competitor for calcium [223]. The growth of apatite crystals is impeded by the Mg^{2+} ions, because they block the active growth sites and interrupt the crystal structure [224].

Organic compounds present in the mitochondrial matrix have a similar effect on phosphate crystals formation. Namely, the nucleotide diphosphates and triphosphates, including ADP and ATP, as well as low molecular weight metabolites containing two attached ester phosphate groups such as citrate also have the ability to inhibit the crystallization of calcium phosphates and may function as in vivo mineralization inhibitors [225,226]. Adsorption of organic compounds on the newly-formed nuclei results in a small crystal size and could be a factor in producing less crystalline phase [198]. Although apatite is usually considered a crystallographic structure, we use the term amorphous carbonated apatite to distinguish it from the amorphous calcium phosphate with the formula $Ca_xH_y(PO_4)_z$.

A characteristic property of apatite compounds is the ability to incorporate ions and individual molecules during its synthesis. One of the most interesting examples is the molecular oxygen, which forms inside the structure through the decomposition of unstable precursor: peroxide ions and hydrogen peroxide molecules trapped in the apatite structure [227]. Mitochondria are involved in the cell's response to oxidative stress, hence the ability of apatite to entrap H_2O_2 and convert it to O_2 might have had a crucial role in the evolution of mitochondria, especially in the early stages before the ability to create enzyme glutathione peroxidase has been developed.

Apatite can incorporate a large variety of foreign ions, including sodium, potassium, strontium, barium, manganese, zinc, iron, cobalt, nickel and even rare earth elements (see [228] for a detailed review). This property is used by a few organisms to eliminate toxic elements and survive in polluted environments, e.g., in gastropods *Littorina Littorea*, apatites allow the accumulation and biochemical deactivation of excess quantities of cadmium and other potentially harmful cations [229,230]. Small organic molecules, such as formate ions and glycine molecules, can also be incorporated in apatites [202,203]. It has been shown that amorphous granular aggregates forming in mitochondria include glycoproteins in their structure [231].

5. Summary and Conclusions

The solubility calculations for calcium and magnesium salts, including carbonates and phosphates, led to the identification of four salts that may form in the mitochondrial matrix: monohydrocalcite, nesquehonite, carbonated apatite and babierrite. Further analysis of inhibition properties of several inorganic and organic ions present in the mitochondrial matrix led to a conclusion, that non-stoichiometric magnesium-rich carbonated apatite precipitates in the matrix under physiological conditions.

The formation of crystalline apatites is inhibited by the presence of magnesium ions, ATP, ADP and citrates. Therefore, amorphous and low-crystalline carbonated apatite is precipitated. Low crystallinity of the precipitated material can be attributed to numerous ionic substitutions in the apatite structure. The general chemical formula of this salt can be expressed as:

$$(Ca,Mg)_{10-x}(PO_4)_{6-x}(HPO_4,CO_3)_x(OH,\frac{1}{2}CO_3,Cl)_{2-x} \text{ with } 0 < x < 2. \tag{7}$$

Numerous ions and molecules can also be incorporated in the structure of apatite, which leads to a further decrease of the crystallinity of precipitated material. The ability to incorporate hydrogen peroxide and decompose it to molecular oxygen might have an auxiliary role in the response to oxidative stress.

Precipitation and dissolution of carbonated apatite inside the matrix, affects the mitochondrial ionic concentrations, and therefore indirectly influences the action of several protein entities in the inner mitochondrial membrane including e.g., ATP synthase, phosphate translocase, calcium uniporter (MCU), mitoK$_{ATP}$ and mitoK$_{Ca}$ channels as well as the K$^+$/H$^+$, Na$^+$/H$^+$ and Ca^{2+}/3Na$^+$ exchangers.

When the oxidative phosphorylation chain is active, the membrane potential on the inner mitochondrial membrane is highly negative, which causes the influx of calcium to the matrix through MCU and leads to the precipitation of carbonated apatite. The dominant forms of carbonic and phosphoric acid in the mitochondrial pH range are HCO$_3^-$ and HPO$_4^{2-}$, respectively. Hence, precipitation of carbonated apatite leads to the production of protons, e.g.,:

$$\begin{aligned}10Ca^{2+} + HCO_3^- + 6HPO_4^{2-} &\to Ca_{10}(PO_4)_6CO_3 + 7H^+ \\ (10-x)Ca^{2+} + xHCO_3^- + (6-x)HPO_4^{2-} + (2-x)H_2O &\leftrightarrow \\ Ca_{10-x}(PO_4)_{6-x}(CO_3)_x(OH)_{2-x} + (8-x)H^+ & \end{aligned} \tag{8}$$

These protons are then pumped outside the mitochondrial matrix by the oxidative phosphorylation chain, which allows the production of ATP.

In anaerobic conditions, i.e., during hypoxia, the oxidative phosphorylation chain stops. Hypoxia can be caused e.g., by ischemia (low blood flow) or carbonate monoxide poisoning (oxygen in hemoglobin replaced by CO). There are three phases of ATP production during ischemia [232]. At first, up to 16 min ATP is produced due to glycolysis cytoplasm acidifies but there is no increase in sodium (5 mM) or calcium concentration (140 nM) in the cytoplasm, i.e., Na-K-ATPase of the cytoplasmic membrane is fully active. In the second phase, as hypothesised by Dołowy [30], during hypoxia, the proton motive force is replaced by the calcium-motive force caused by the dissolution of calcium phosphates, efflux of calcium ions and alkalization of the matrix due to the following processes:

$$\begin{aligned}Ca_{10}(PO_4)_6CO_3 + 7H_2O &\to 10Ca^{2+}\uparrow + CO_2\uparrow + 6HPO_4^{2-} + 8OH^- \\ Ca_{10-x}(PO_4)_{6-x}(CO_3)_x(OH)_{2-x} + 6H_2O &\to (10-x)Ca^{2+}\uparrow + xCO_2\uparrow + (6-x)HPO_4^{2-} + 8OH^-\end{aligned} \tag{9}$$

where ↑ represents ions or molecules leaving the mitochondrial matrix. Phosphate ions forming due to the dissolution of carbonated apatite and lower ATP/ADP ratio are the two factors responsible for lowering the free energy of the ATP synthesis, hence allowing its production in hypoxic conditions. During the second phase which lasts up to 30 min, sodium concentration in the cytoplasm is slowly rising to 30 mM and calcium concentration to 500 nM level. The mitochondrion can survive, as long as it can keep the pH of the matrix constant by the precipitation and dissolution of carbonated apatite. Then during the third phase, when all available salt dissolves, further influx of protons leads to the acidification of the matrix, hence lowering the proton motive force below the level necessary for ATP production. When matrix acidifies Na/H and K/H exchangers cannot prevent influx of osmotically active cations into the matrix and mitochondria swells. Low ATP levels in the cytoplasm lead to opening of the mitoK$_{ATP}$ channel. Upon reperfusion sharp increase in mitochondrial electric potential leads to further influx of osmotically active cations into the mitochondrial matrix and formation of the permeability transition pore [233,234].

The build-up of amorphous carbonated apatite should increase the maximum time of hypoxia that the cell can survive. It explains the increased life-time observed in the heart cells preconditioned by short increases of the calcium concentration [235,236]. Calcium preconditioning increases the cytoplasmic calcium ion concentration. Therefore Ca^{2+} from the cytoplasm is more rapidly transported into the matrix, which leads to the precipitation of carbonated apatite. A better understanding of the relationship between calcium storage and the mechanisms of ischemic preconditioning may lead to the development of new stroke-protective drugs.

Author Contributions: J.J.J. conceived, designed and performed all the calculations, analyzed the data, reviewed the available literature, wrote the paper, and supervised the overall research. R.F., K.D. and A.L. discussed the research. A.L. acquired funding. All authors have read and agreed to the published version of the manuscript.

Funding: National Science Centre (NCN, Poland) financial support via research grant no. 2014/15/B/ST5/02185 is acknowledged.

Conflicts of Interest: The authors declare no conflict of interest.

References

1. Santo-Domingo, J.; Demaurex, N. The renaissance of mitochondrial pH. *J. Gen. Physiol.* **2012**, *139*, 415–423. [CrossRef] [PubMed]
2. Griffiths, T.; Evans, M.C.; Meldrum, B.S. Intracellular sites of early calcium accumulation in the rat hippocampus during status epilepticus. *Neurosci. Lett.* **1982**, *30*, 329–334. [CrossRef]
3. Erecińska, M.; Silver, I.A. Relationships between ions and energy metabolism: Celebral calcium movements during ischemia and subsequent recovery. *Can. J. Physiol. Pharmacol.* **1992**, *70*, S190–S193. [CrossRef] [PubMed]
4. Zaidan, E.; Sims, N.R. The calcium content of mitochondria from brain subregions following short-term forebrain ischemia and recirculation in the rat. *J. Neurochem.* **1994**, *63*, 1812–1819. [CrossRef] [PubMed]
5. Fineman, I.; Hovda, D.A.; Smith, M.; Yoshino, A.; Becker, D.P. Concussive brain injury is associated with a prolonged accumulation of calcium: A 45Ca autoradiographic study. *Brain Res.* **1993**, *624*, 94–102. [CrossRef]
6. Nelson, D.L; Cox, M.M. *Lehninger Principles of Biochemistry*, 6th ed.; Freeman/Worth: New York, USA, 2012; pp. 690–748.
7. Lehninger, A.L.; Rossi, C.S.; Greenawalt, J.W. Respiration-dependent accumulation of inorganic phosphate and Ca ions by rat liver mitochondria. *Biochem. Biophys. Res. Commun.* **1963**, *10*, 444–448. [CrossRef]
8. Lehninger, A.L. Mitochondria and calcium ion transport. *Biochem. J.* **1970**, *119*, 129–138. [CrossRef] [PubMed]
9. Rossi, C.S.; Lehninger, A.L. Stoichiometric relationships between accumulation of ions by mitochondria and the energy-coupling sites in the respiratory chain. *Biochem. Z.* **1963**, *338*, 698–713.
10. Rossi, C.S.; Lehninger, A.L. Stoichiometry of respiratory stimulation, accumulation of Ca^{2+} and phosphate and oxidative phosphorylation in rat liver mitochondria. *J. Biol. Chem.* **1964**, *239*, 3971–3980.
11. Lehninger, A.L.; Carafoli, E.; Rossi, C.S. Energy-linked ion movements in mitochondrial systems. *Adv. Enzymol. Relat. Areas Mol. Biol.* **1967**, *29*, 259–320.
12. Brookes, P.S.; Yoon, Y.; Robotham, J.L.; Anders, M.W.; Sheu, S. Calcium, ATP, and ROS: A mitochondrial love-hate triangle. *Am. J. Physiol.* **2004**, *287*, 817–833. [CrossRef] [PubMed]
13. Clapham, D.E. Calcium signaling. *Cell* **2007**, *131*, 1047–1058. [CrossRef] [PubMed]
14. McCormack, J.G.; Denton, R.M. The role of mitochondrial Ca^{2+} transport and matrix Ca^{2+} in signal transduction in mammalian tissues. *Biochim. Biophys. Acta* **1990**, *1018*, 287–291. [CrossRef]
15. Gonzales, F.; Karnovsky, M.J. Electron microscopy of osteoblasts in healing fractures of rat bone. *J. Biophys. Biochem. Cytol.* **1961**, *9*, 299–316. [CrossRef] [PubMed]
16. Sutfin, L.V.; Holtrop, M.E.; Ogilvie, R.E. Microanalysis of individual mitochondrial granules with diameters less than 1000 angstroms. *Science* **1971**, *174*, 947–949. [CrossRef] [PubMed]
17. Martin, J.H.; Matthews, J.L. Mitochondrial granules in chondrocytes. *Calcif. Tissue Res.* **1969**, *3*, 184–193. [CrossRef]
18. Landis, W.J.; Paine, M.C.; Glimcher, M.J. Use of acrolein vapors for the anhydrous preparation of bone tissue for electron microscopy. *J. Ultrastruct. Res.* **1980**, *70*, 171–180. [CrossRef]
19. Matthews, J.L. Ultrastructure of calcifying tissues. *Am. J. Anat.* **1970**, *129*, 450–457. [CrossRef]

20. Martin, J.H.; Matthews, J.L. Mitochondrial granules in chondrocytes, osteoblasts and osteocytes: An ultrastructural and microincineration study. *Clin. Orthop. Relat. Res.* **1970**, *68*, 273–278. [CrossRef]
21. Matthews, J.L.; Martin, J.H.; Sampson, H.W.; Kunin, A.S.; Roan, J.H. Mitochondrial granules in the normal and rachitic rat epiphysis. *Calcif. Tissue Res.* **1970**, *5*, 91–99. [CrossRef]
22. Landis, W.J.; Glimcher, M.J. Electron optical and analytical observations of rat growth plate cartilage prepared by ultracryomicrotomy: The failure to detect a mineral phase in matrix vesicles and the identification of heterodispersed particles as the initial solid phase of calcium phosphate deposited in the extracellular matrix. *J. Ultrastruct. Res.* **1982**, *78*, 227–268. [PubMed]
23. Gay, C.; Schraer, H. Frozen thin-sections of rapidly forming bone: Bone cell ultrastructure. *Calcif. Tissue Res.* **1975**, *19*, 39–49. [CrossRef]
24. Landis, W.J.; Glimcher, M.J. Electron diffraction and electron probe microanalysis of the mineral phase of bone tissue prepared by anhydrous techniques. *J. Ultrastruct. Res.* **1978**, *63*, 188–223. [CrossRef]
25. Landis, W.J.; Hauschka, B.T.; Rogerson, C.A.; Glimcher, M.J. Electron microscopic observations of bone tissue prepared by ultracryomicrotomy. *J. Ultrastruct. Res.* **1977**, *59*, 185–206. [CrossRef]
26. Nicholls, D.G.; Chalmers, S. The integration of mitochondrial calcium transport and storage. *J. Bioenerg. Biomembr.* **2004**, *36*, 277–281. [CrossRef]
27. Kristian, T.; Pivovarova, N.B.; Fiskum, G.; Andrews, S.B. Calcium-induced precipitate formation in brain mitochondria: Composition, calcium capacity, and retention. *J. Neurochem.* **2007**, *102*, 1346–1356. [CrossRef] [PubMed]
28. Thomas, R.S.; Greenawalt, J.W. Microincineration, electron microscopy, and electron diffraction of calcium phosphateloaded mitochondria. *J. Cell Biol.* **1968**, *39*, 55–76. [CrossRef]
29. Chalmers, S.; Nichols, D.G. The relationship between free and total calcium concentration in the matrix of liver and brain mitochondria. *J. Biol. Chem.* **2003**, *278*, 19062–19070. [CrossRef] [PubMed]
30. Dołowy, K. Calcium phosphate buffer formed in the mitochondrial matrix during preconditioning supports ΔpH formation and ischemic ATP production and prolongs cell survival—A hypothesis. *Mitochondrion* **2019**, *47*, 210–217. [CrossRef]
31. Bal, W.; Kurowska, E.; Maret, W. The Final Frontier of pH and the Undiscovered Country Beyond. *PLoS ONE* **2012**, *7*, e45832. [CrossRef]
32. Bose, S.; French, S.; Evans, F.J.; Joubert, F.; Balaban, R.S. Metabolic network control of oxidative phosphorylation. Multiple roles of inorganic phosphate. *J. Biol. Chem.* **2003**, *278*, 39155–39165. [CrossRef] [PubMed]
33. Baysal, K.; Brierley, G.P.; Novgorodov, S.; Jung, D.W. Regulation of the mitochondrial Na^+/Ca^{2+} antiport by matrix pH. *Arch. Biochem. Biophys.* **1991**, *291*, 383–389. [CrossRef]
34. Balut, C.; vandeVen, M.; Despa, S.; Lambrichts, I.; Ameloot, M.; Steels, P.; Smets, I. Measurement of cytosolic and mitochondrial pH in living cells during reversible metabolic inhibition. *Kidney Int.* **2008**, *73*, 226–232. [CrossRef] [PubMed]
35. Takahashi, A.; Zhang, Y.; Centonze, V.E.; Herman, B. Measurement of Mitochondrial pH In Situ. *BioTechniques* **2001**, *30*, 804–815. [CrossRef]
36. Abad, M.F.C.; Di Benedetto, G.; Magalhães, P.J.; Filippin, L.; Pozzan, T. Mitochondrial pH monitored by a new engineered green fluorescent protein mutant. *J. Biol. Chem.* **2003**, *279*, 11521–11529. [CrossRef]
37. Matsuyama, S.; Llopis, J.; Deveraux, Q.L.; Tsien, R.Y.; Reed, J.C. Changes in intramitochondrial and cytosolic pH: Early events that modulate caspase activation during apoptosis. *Nature Cell Biol.* **2000**, *2*, 318–325. [CrossRef]
38. Poburko, D.; Santo-Domingo, J.; Demaurex, N. Dynamic regulation of the mitochondrial proton gradient during cytosolic calcium elevations. *J. Biol. Chem.* **2011**, *286*, 11672–11684. [CrossRef]
39. Hutson, S. M. pH regulation of mitochondrial branched chain α-keto acid transport and oxidation in rat heart mitochondria. *J. Biol. Chem.* **1987**, *262*, 9629–9635.
40. Ivannikov, M.V.; Macleod, G.T. Mitochondrial free Ca^{2+} levels and their effects on energy metabolism in Drosophila motor nerve terminals. *Biophys. J.* **2013**, *104*, 2353–2361. [CrossRef]
41. Al-Nassar, I.; Crompton, M. The entrapment of the Ca^{2+} indicator arsenazo III in the matrix space of rat liver mitochondria by permeabilization and resealing. Na^+-dependent and -independent effluxes of Ca^{2+} in arsenazo III-loaded mitochondria. *Biochem. J.* **1986**, *239*, 31–40. [CrossRef]

42. Samson, E.; Lemaire, G.; Marchand, J.; Beaudoin, J.J. Modeling chemical activity effects in strong ionic solutions. *Comp. Mat. Sci.* **1999**, *15*, 285–294. [CrossRef]
43. Jung, D.W.; Davis, M.H.; Brierley, G.P. Estimation of matrix pH in isolated heart mitochondria using a fluorescent probe. *Anal. Biochem.* **1989**, *178*, 348–354. [CrossRef]
44. Addanki, S.; Cahill, F.D.; Sotos, J.F. Determination of intramitochondrial pH and intramitochondrial-extramitochondrial pH gradient of isolated heart mitochondria by the use of 5,5-dimethyl-,4- oxazolidinedione. *J. Biol. Chem.* **1968**, *243*, 2337–2348. [PubMed]
45. Porcelli, A.M.; Ghelli, A.; Zanna, C.; Pinton, P.; Rizzuto, R.; Rugolo, M. pH difference across the outer mitochondrial membrane measured with a green fluorescent protein mutant. *Biochem. Biophys. Res. Commun.* **2005**, *326*, 799–804. [CrossRef] [PubMed]
46. Llopis, J.; McCaffery, J.M.; Miyawaki, A.; Farquhar, M.G.; Tsien, R.Y. Measurement of cytosolic, mitochondrial, and Golgi pH in single living cells with green fluorescent proteins. *Proc. Natl. Acad. Sci.* **1998**, *95*, 6803–6808. [CrossRef] [PubMed]
47. Greenbaum, N.L.; Wilson, D.F. Role of intramitochondrial pH in the energetics and regulation of mitochondrial oxidative phosphorylation. *Biochim. Biophys. Acta Bioenerg.* **1991**, *1058*, 113–120. [CrossRef]
48. Schoolwerth, A.C.; LaNoue, K.F.; Hoover, W.J. Effect of pH on glutamate efflux from rat kidney mitochondria. *Am. J. Physiol. Renal Physiol.* **1984**, *246*, F266–F271. [CrossRef]
49. Baron, S. Role of mitochondrial Na+ concentration, measured by CoroNa Red, in the protection of metabolically inhibited MDCK cells. *J. Am. Soc. Nephrol.* **2005**, *16*, 3490–3497. [CrossRef]
50. Jung, D.W.; Apel, L.M.; Brierley, G.P. Transmembrane gradients of free Na+ in isolated heart mitochondria estimated using a fluorescent probe. *Am. J. Physiol.* **1992**, *262*, C1047–C1055. [CrossRef]
51. Donoso, P.; Mill, J.G.; O'Neill, S.C.; Eisner, D.A. Fluorescence measurements of cytoplasmic and mitochondrial sodium concentration in rat ventricular myocytes. *J. Physiol.* **1992**, *448*, 493–509. [CrossRef]
52. Augustynek, B.; Wrzosek, A.; Koprowski, P.; Kiełbasa, A.; Bednarczyk, P.; Łukasiak, A.; Dołowy, K.; Szewczyk, A. What we don't know about mitochondrial potassium channels? *Postępy Biochemii* **2016**, *62*, 189–198. [PubMed]
53. Zoeteweij, J.P.; van de Water, B.; de Bont, H.J.; Nagelkerke, J.F. Mitochondrial K$^+$ as modulator of Ca^{2+}-dependent cytotoxicity in hepatocytes. Novel application of the K$^+$-sensitive dye PBFI (K$^+$-binding benzofuran isophthalate) to assess free mitochondrial K$^+$ concentrations. *Biochem. J.* **1994**, *299*, 539–543. [CrossRef] [PubMed]
54. Costa, A.D.T.; Quinlan, C.L.; Andrukhiv, A.; West, I.C.; Jabůrek, M.; Garlid, K.D. The direct physiological effects of mitoKATP opening on heart mitochondria. *Am. J. Physiol. Heart Circ. Physiol.* **2006**, *290*, H406–H415. [CrossRef] [PubMed]
55. Dordick, R.S.; Brierley, G.P.; Garlid, K.D. On the mechanism of A23187-induced potassium efflux in rat liver mitochondria. *J. Biol. Chem.* **1980**, *255*, 10299–10305. [PubMed]
56. Corkey, B.E.; Duszynski, J.; Rich, T.L.; Matschinsky, B.; Williamson, J.R. Regulation of free and bound magnesium in rat hepatocyte and isolated mitochondria. *J. Biol. Chem.* **1986**, *261*, 2567–2574. [PubMed]
57. Jung, D.W.; Brierley, G.P. Matrix magnesium and the permeability of heart mitochondria to potassium ion. *J. Biol. Chem.* **1986**, *261*, 6408–6415. [PubMed]
58. Gout, E.; Rebeille, F.; Douce, R.; Bligny, R. Interplay of Mg^{2+}, ADP, and ATP in the cytosol and mitochondria: Unravelling the role of Mg^{2+} in cell respiration. *Proc. Natl. Acad. Sci.* **2014**, *111*, E4560–E4567. [CrossRef]
59. Jung, D.W.; Apel, L.; Brierley, G.P. Matrix free magnesium changes with metabolic state in isolated heart mitochondria. *Biochemistry* **1990**, *29*, 4121–4128. [CrossRef]
60. Rutter, G.A.; Osbaldeston, N.J.; McCormack, J.G.; Denton, R.M. Measurement of matrix free Mg^{2+} concentration in rat heart mitochondria by using entrapped fluorescent probes. *Biochem. J.* **1990**, *271*, 627–634. [CrossRef]
61. Arnaudeau, S.; Kelley, W.L.; Walsh, J.V.; Demaurex, N. Mitochondria recycle Ca^{2+} to the endoplasmic reticulum and prevent the depletion of neighboring endoplasmic reticulum regions. *J. Biol. Chem.* **2001**, *276*, 29430–29439. [CrossRef]
62. Brandenburger, Y.; Kennedy, E.D.; Python, C.P.; Rossier, M.F.; Vallotton, M.B.; Wollheim, C.B.; Capponi, A.M. Possible role for mitochondrial calcium in angiotensin II- and potassium-stimulated steroidogenesis in bovine adrenal glomerulosa cells. *Endocrinology* **1996**, *137*, 5544–5551. [CrossRef] [PubMed]

63. Schreur, J.H.; Figueredo, V.M.; Miyamae, M.; Shames, D.M.; Baker, A.J.; Camacho, S.A. Cytosolic and mitochondrial [Ca^{2+}] in whole hearts using indo-1 acetoxymethyl ester: effects of high extracellular Ca^{2+}. *Biophys. J.* **1996**, *70*, 2571–2580. [CrossRef]
64. Miyata, H.; Silverman, H.S.; Sollott, S.J.; Lakatta, E.G.; Stern, M.D.; Hansford, R.G. Measurement of mitochondrial free Ca^{2+} concentration in living single rat cardiac myocytes. *Am. J. Physiol.* **1991**, *261*, H1123–H1134. [CrossRef]
65. Moreno-Sanchez, R.; Hansford, R.G. Dependence of cardiac mitochondrial pyruvate dehydrogenase activity on intramitochondrial free Ca^{2+} concentration. *Biochem. J.* **1988**, *256*, 403–412. [CrossRef] [PubMed]
66. Allen, S.P.; Stone, D.; McCormack, J.G. The loading of fura-2 into mitochondria in the intact perfused rat heart and its use to estimate matrix Ca^{2+} under various conditions. *J. Mol. Cell. Cardiol.* **1992**, *24*, 765–773. [CrossRef]
67. Lukács, G.L.; Kapus, A. Measurement of the matrix free Ca^{2+} concentration in heart mitochondria by entrapped fura-2 and quin2. *Biochem. J.* **1987**, *248*, 609–613. [CrossRef]
68. Davis, M.H.; Altschuld, R.A.; Jung, D.W.; Brierley, G.P. Estimation of intramitochondrial pCa and pH by fura-2 and 2,7 biscarboxyethyl-5(6)-carboxyfluorescein (BCECF) fluorescence. *Biochem. Biophys. Res. Commun.* **1987**, *149*, 40–45. [CrossRef]
69. Jahn, S.C.; Rowland-Faux, L.; Stacpoole, P.W.; James, M.O. Chloride concentrations in human hepatic cytosol and mitochondria are a function of age. *Biochem. Biophys. Res. Commun.* **2015**, *459*, 463–468. [CrossRef]
70. Kielland, J. Individual Activity Coefficients of Ions in Aqueous Solutions. *J. Am. Chem. Soc.* **1937**, *59*, 1675–1678. [CrossRef]
71. Arthurs, G.J.; Sudhakar, M. Carbon dioxide transport. *Contin. Educ. Anaesth. Critical Care Pain* **2005**, *5*, 207–210. [CrossRef]
72. Sun, X.-G.; Hansen, J.E.; Stringer, W.W.; Ting, H.; Wasserman, K. Carbon dioxide pressure-concentration relationship in arterial and mixed venous blood during exercise. *J. Appl. Physiol.* **2001**, *90*, 1798–1810. [CrossRef] [PubMed]
73. Henry, W. Experiments on the quantity of gases absorbed by water, at different temperatures, and under different pressures. *Phil. Trans. R. Soc. Lond.* **1803**, *93*, 29–274.
74. Sander, R. Compilation of Henry's law constants (version 4.0) for water as solvent. *Atmos. Chem. Phys.* **2015**, *15*, 4399–4981. [CrossRef]
75. Plummer, L.N.; Busenberg, E. The solubilities of calcite, aragonite and vaterite in CO$_2$-H$_2$O solutions between 0 and 90 °C, and an evaluation of the aqueous model for the system CaCO$_3$-CO$_2$-H$_2$O. *Geochim. Cosmochim. Acta* **1982**, *46*, 1011–1040. [CrossRef]
76. Dąbrowska, S.; Migdalski, J.; Lewenstam, A. A Breakthrough Application of a Cross-Linked Polystyrene Anion-Exchange Membrane for a Hydrogencarbonate Ion-Selective Electrode. *Sensors* **2019**, *19*, 1268. [CrossRef]
77. Harned, H.S.; Davis, R.D., Jr. The ionization constant of carbonic acid in water and the solubility of carbon dioxide in water and aqueous salt solutions from 0 to 50 °C. *J. Am. Chem. Soc.* **1943**, *65*, 2030–2037. [CrossRef]
78. Harned, H.S.; Bonner, F.T. The first ionization constant of carbonic acid in aqueous solutions of sodium chloride. *J. Am. Chem. Soc.* **1945**, *67*, 1026–1031. [CrossRef]
79. Harned, H.S.; Scholes, S.R. The ionization constant of HCO3- from 0 to 50 °C. *J. Am. Chem. Soc.* **1941**, *63*, 1706–1709. [CrossRef]
80. Ryzhenko, B.N. Determination of dissociation constants of carbonic acid and the degree of hydrolysis of the CO:- and HCO; ions in solutions of alkali carbonates and bicarbonates at elevated temperatures. *Geochemistry* **1963**, *2*, 151–164.
81. Millero, F.J. The thermodynamics of the carbonate system in seawater. *Geochim. Cosmochim. Acta* **1979**, *43*, 1651–1661. [CrossRef]
82. Prieto, F.J.M.; Millero, F.J. The determination of pK1+pK2 in seawater as a function of temperature and salinity. *Geochim. Cosmochim. Acta* **2002**, *66*, 2529–2540.
83. Millero, F. J.; Graham, T.B.; Huang, F.; Bustos-Serrano, H.; Pierrot, D. Dissociation constants of carbonic acid in seawater as a function of salinity and temperature. *Marine Chem.* **2006**, *100*, 80–94. [CrossRef]
84. Flear, C.T.G.; Covington, A.K.; Stoddart, J.C. Bicarbonate or CO$_2$? *Arch. Int. Med.* **1984**, *144*, 2285–2287. [CrossRef]

85. Flear, C.T.G.; Roberts, S.W.; Hayes, S.; Stoddart, J.C.; Covington, A.K. pK_1' and bicarbonate concentration in plasma. *Clin. Chem.* **1987**, *33*, 13–20. [CrossRef] [PubMed]
86. Tibi, L.; Bhattacharya, S.S.; Flear, C.T.G. Variability of pK_1' of human plasma. *Clin. Chim. Acta* **1982**, *121*, 15–31. [CrossRef]
87. Natelson, S.; Nobel, D. Effect of the variation of pK' of the Henderson-Hasselbalch equation on values obtained for total CO2 calculated from pCO2 and pH values. *Clin. Chem.* **1977**, *23*, 767–769.
88. Masters, P.; Blackburn, M.E.C.; Henderson, M.J.; Barrett, J.F.R.; Dear, P.R.F. Determination of plasma bicarbonate of neonates in intensive care. *Clin. Chem.* **1988**, *34*, 1483–1485. [CrossRef]
89. Rosan, R.; Enlander, D.; Ellis, S. Unpredictable error in calculated bicarbonate homeostasis during pediatric intensive care: The delusion of the fixed pK'. *Clin. Chem.* **1983**, *29*, 69–73. [CrossRef]
90. Kost, G.J.; Trent, J.K.T.; Saeed, D. Indications for measurement of total carbon dioxide in arterial blood. *Clin. Chem.* **1988**, *34*, 1650–1652. [CrossRef]
91. Fresenius, R. Ueber die Löslichkoitsverhältnisse voneinigen bei der quantitativen Analyse als Bestimmungsformen, etc., dienenden Niederschlägen. *Ann. Chem. Pharm.* **1846**, *59*, 117–128. [CrossRef]
92. Lassaigne, J.L. Löslichkeit einiger kohlensauren Salze in kohlensaurem Wasser. *Ann. Chem. Pharm.* **1848**, *68*, 253–254.
93. De Visscher, A.; Vanderdeelen, J. Estimation of the Solubility Constant of Calcite, Aragonite, and Vaterite at 25 °C Based on Primary Data Using the Pitzer Ion Interaction Approach. *Monatshefte für Chemie* **2003**, *134*, 769–775. [CrossRef]
94. Clarkson, J.R.; Price, T.J.; Adams, C.J. Role of metastable phases in the spontaneous precipitation of calcium carbonate. *J. Chem. Sot. Faraday Trans.* **1992**, *88*, 243–249. [CrossRef]
95. Brečević, L.; Nielsen, A.E. Solubility of Calcium Carbonate Hexahydrate. *Acta Chim. Scandinav.* **1993**, *47*, 668–673. [CrossRef]
96. Kralj, D.; Brečević, L. Dissolution kinetics and solubility of calcium carbonate monohydrate. *Colloids Surf. A: Physicochem. Eng. Aspects* **1995**, *96*, 287–293. [CrossRef]
97. Hull, H.; Turnbull, A.G. A thermochemical study of monohydrocalcite. *Geochim. Cosmochim. Acta* **1973**, *37*, 685–694. [CrossRef]
98. Brečević, L.; Nielsen, A.E. Solubility of amorphous calcium carbonate. *J. Cryst. Growth* **1989**, *98*, 504–510. [CrossRef]
99. Canterford, J.H.; Tsambourakis, G.; Lambert, B. Some observations on the properties of dypingite, $Mg_5(CO_3)_4(OH)_2 \cdot 5H_2O$, and related minerals. *Mineral. Mag.* **1984**, *48*, 437–442. [CrossRef]
100. Hänchen, M.; Prigiobbe, V.; Baciocchi, R.; Mazzotti, M. Precipitation in the Mg-carbonate system—Effects of temperature and CO_2 pressure. *Chem. Eng. Sci.* **2008**, *63*, 1012–1028. [CrossRef]
101. Ropp, R.C. *Encyclopedia of the Alkaline Earth Compounds*, 1st ed.; Elsevier Science Ltd.: Oxford, UK, 2013; pp. 360–362.
102. Rheinheimer, V.; Unluer, C.; Liu, J.; Ruan, S.; Pan, J.; Monteiro, P. XPS Study on the Stability and Transformation of Hydrate and Carbonate Phases within MgO Systems. *Materials* **2017**, *10*, 75. [CrossRef]
103. Mineralogy Database. Available online: https://www.mindat.org/ (accessed on 4 February 2020).
104. Nashar, B. Barringtonite—A new hydrous magnesium carbonate from Barrington Tops, New South Wales, Australia. *Mineral. Mag. J. Mineral. Soc.* **1965**, *34*, 370–372. [CrossRef]
105. Fitzpatrick, J.J. Pokrovskite: Its possible relationship to mcguinnessite and the problem of excess water. In Proceedings of the 14th Meeting International Mineralogical Association, Abstracts, Stanford, CA, USA, 13–18 July 1986; p. 101.
106. Gunter, J.R.; Oswald, H.R. Crystal structure of $Mg_2(OH)_2(CO_3)$, deduced from the topotactic thermal decomposition of artinite. *J. Solid State Chem.* **1977**, *21*, 211–215. [CrossRef]
107. Hopkinson, L.; Kristova, P.; Rutt, K.; Cressey, G. Phase transitions in the system $MgO–CO_2–H_2O$ during CO_2 degassing of Mg-bearing solutions. *Geochim. Cosmochim. Acta* **2012**, *76*, 1–13. [CrossRef]
108. Di Tommaso, D.; De Leeuw, N.H. Structure and dynamics of the hydrated magnesium ion and of the solvated magnesium carbonates: Insights from first principles simulations. *Phys. Chem. Chem. Phys.* **2010**, *12*, 894–901. [CrossRef] [PubMed]
109. Saldi, G.D.; Jordan, G.; Schott, J.; Oelkers, E.H. Magnesite growth rates as a function of temperature and saturation state. *Geochim. Cosmochim. Acta* **2009**, *73*, 56465657. [CrossRef]

110. Saldi, G.D.; Schott, J.; Pokrovsky, O.S.; Gautier, Q.; Oelkers, E.H. An experimental study of magnesite precipitation rates at neutral to alkaline conditions and 100–200 °C as a function of pH, aqueous solution composition and chemical affinity. *Geochim. Cosmochim. Acta* **2012**, *83*, 93109. [CrossRef]

111. Sayles, F.L.; Fyfe, W.S. The crystallization of magnesite from aqueous solution. *Geochim. Cosmochim. Acta* **1973**, *37*, 87–99. [CrossRef]

112. Dell, R.; Weller, S.W. The thermal decomposition of nesquehonite $MgCO_3 \cdot H_2O$ and magnesium ammonium carbonate $MgCO_3 \cdot (NH_4)_2CO_3 \cdot 4H_2O$. *Trans. Faraday Soc.* **1959**, *55*, 2203–2220. [CrossRef]

113. Ming, D.W.; Franklin, W.T. Synthesis and Characterization of Lansfordite and Nesquehonite. *Soil Sci. Soc. Am. J.* **1985**, *49*, 1303–1308. [CrossRef]

114. Davies, P.J.; Bubela, B. The transformation of nesquehonite into hydromagnesite. *Chem. Geol.* **1973**, *12*, 289–300. [CrossRef]

115. Kloprogge, J.T.; Martens, W.N.; Nothdurft, L.; Duong, L.V.; Webb, G.E. Low temperature synthesis and characterization of nesquehonite. *J. Mater. Sci. Lett.* **2003**, *22*, 825–829. [CrossRef]

116. Zhang, Z.P.; Zheng, Y.J.; Ni, Y.W.; Liu, Z.M.; Chen, J.P.; Liang, X.M. Temperature- and pH-dependent morphology and FT-IR analysis of magnesium carbonate hydrates. *J. Phys. Chem. B* **2006**, *110*, 12969–12973. [CrossRef] [PubMed]

117. Fernandez, A.I.; Chimenos, J.M.; Segarra, M.; Fernandez, M.A.; Espiell, F. Procedure to obtain hydromagnesite from a MgO-containing residue. Kinetic study. *Ind. Eng. Chem. Res.* **2000**, *39*, 3653–3658. [CrossRef]

118. Kline, W. D. The solubility of magnesium carbonate (nesquehonite) in water at 25 °C and pressures of carbon dioxide up to one atmosphere. *Am. Chem. Soc. J.* **1929**, *51*, 2093–2097. [CrossRef]

119. Langmuir, D. Stability of Carbonates in the System $MgO-CO_2-H_2O$. *J. Geol.* **1965**, *73*, 730–754. [CrossRef]

120. Harrison, A.L.; Mavromatis, V.; Oelkers, E.H.; Bénézeth, P. Solubility of the hydrated Mg-carbonates nesquehonite and dypingite from 5 to 35 °C: Implications for CO_2 storage and the relative stability of Mg-carbonates. *Chem. Geol.* **2019**, *504*, 123–135. [CrossRef]

121. Johnston, J. The solubility-product constant of calcium and magnesium carbonates. *J. Am. Chem. Soc.* **1915**, *37*, 2001–2020. [CrossRef]

122. Hemingway, B.S.; Robie, R.A. A calorimetric determination of the standard enthalpies of formation of huntite, $CaMg(CO_3)_4$, and artinite, $Mg_2(OH)_2CO_3 \cdot 3H_2O$, and their standard Gibbs free energies of formation. *J. Res. U.S. Geol. Surv.* **1972**, *1*, 535–541.

123. Cheng, W.; Li, Z. Controlled Supersaturation Precipitation of Hydromagnesite for the $MgCl_2-Na_2CO_3$ System at Elevated Temperatures: Chemical Modeling and Experiment. *Ind. Eng. Chem. Res.* **2010**, *49*, 1964–1974. [CrossRef]

124. Gautier, Q.; Bénézeth, P.; Mavromatis, V.; Schott, J. Hydromagnesite solubility product and growth kinetics in aqueous solution from 25 to 75 °C. *Geochim. Cosmochim. Acta* **2014**, *138*, 1–20. [CrossRef]

125. Raade, G. Dypingite, a new hydrous basic carbonate of magnesium, from Norway. *Am. Mineral.* **1970**, *55*, 1457–1465.

126. Botha, A.; Strydom, C.A. Preparation of a magnesium hydroxy carbonate from magnesium hydroxide. *Hydrometallurgy* **2001**, *62*, 175–183. [CrossRef]

127. Power, I.M.; Wilson, S.A.; Thom, J.M.; Dipple, G.M.; Southam, G. Biologically induced mineralization of dypingite by cyanobacteria from an alkaline wetland near Atlin, British Columbia, Canada. *Geochem. Trans.* **2007**, *8*, 13. [CrossRef] [PubMed]

128. Larsen, M.J. An investigation of the theoretical background for the stability of the calcium-phosphate salts and their mutual conversion in aqueous solutions. *Arch. Oral Biol.* **1986**, *31*, 757–761. [CrossRef]

129. Fernández, E.; Gil, F.J.; Ginebra, M.P.; Driessens, F.C.M.; Planell, J.A.; Best, S.M. Calcium phosphate bone cements for clinical applications Part I: Solution chemistry. *J. Mat. Sci. Mater. Med.* **1999**, *10*, 169–176. [CrossRef]

130. Dorozhkin, S.V. Calcium orthophosphates. *J. Mat. Sci.* **2007**, *42*, 1061–1095. [CrossRef]

131. Dorozhkin, S.V. Calcium orthophosphates. *Biomatter* **2011**, *1*, 121–164. [CrossRef]

132. Sandberger, F. Ueber Isoklas und Kollophan, zwei neue Phosphate. *J. Prakt. Chemie Neue Folge* **1870**, *2*, 125–130. [CrossRef]

133. Ondruš, P.; Veselovský, F.; Hloušek, J.; Skála, R.; Vavřín, I.; Frýda, J.; Čejka, J.; Gabašová, A. Secondary minerals of the Jáchymov (Joachimsthal) ore district. *J. Czech Geol. Soc.* **1997**, *42*, 3–76.

134. Madsen, H.E.L. Ionic Concentration in Calcium Phosphate Solutions, I. Solutions Saturated with Respect to Brushite or Tetracalcium Monohydrogen Phosphate at 37 °C. *Acta Chem. Scand.* **1970**, *24*, 1671–1676. [CrossRef]
135. Sutter, J.R.; McDowell, H.; Brown, W.E. Solubility study of calcium hydrogen phosphate. Ion-pair formation. *Inorg. Chem.* **1971**, *10*, 1638–1643. [CrossRef]
136. Tung, M.S.; Eidelman, N.; Sieck, B.; Brown, W.E. Octacalcium phosphate solubility product from 4 to 37 °C. *J. Res. Nat. Bur. Stand.* **1988**, *93*, 613–624. [CrossRef]
137. Shyu, L.J.; Perrez, L.; Zawacky, S.J.; Heughebaert, J.C.; Nancollas, G.H. The Solubility of Octacalcium Phosphate at 37 °C in the System $Ca(OH)_2$-H_3PO_4-KNO_3-H_2O. *J. Dent. Res.* **1983**, *62*, 398–400. [CrossRef]
138. Christoffersen, M.R.; Christoffersen, J.; Kibalczyc, W. Apparent solubilities of two amorphous calcium phosphates and of octacalcium phosphate in the temperature range 30–42 °C. *J. Cryst. Growth* **1990**, *106*, 349–354. [CrossRef]
139. Ito, A.; Maekawa, K.; Tsutsumi, S.; Ikazaki, F.; Tateishi, T. Solubility product of OH-carbonated hydroxyapatite. *J. Biomed. Mater. Res.* **1997**, *36*, 522–528. [CrossRef]
140. Narasaraju, T.S.B.; Rao, K.K.; Rai, U.S. Determination of solubility products of hydroxylapatite, chlorapatite, and their solid solutions. *Can. J. Chem.* **1979**, *57*, 1919–1922. [CrossRef]
141. Moreno, E.C.; Kresak, M.; Zahradnik, R.T. Physicochemical Aspects of Fluoride-Apatite Systems Relevant to the Study of Dental Caries. *Caries Res.* **1977**, *11*, 142–171. [CrossRef]
142. Schrödter, K.; Bettermann, G.; Staffel, T.; Wahl, F.; Klein, T.; Hofmann, T. Phosphoric Acid and Phosphates. In *Ullmann's Encyclopedia of Industrial Chemistry*, 7th ed.; Wiley-VCH: Weinheim, Germany, 2012; Volume 26, pp. 679–712.
143. Dudenhoefer, R.; Messing, G.L.; Brown, P.W.; Johnson, G.G. Synthesis and characterization of monomagnesium phosphate tetrahydrate. *J. Cryst. Growth* **1992**, *125*, 121–126. [CrossRef]
144. Taylor, A.W.; Frazier, A.W.; Gurney, E.L.; Smith, J.P. Solubility products of di- and trimagnesium phosphates and the dissociation of magnesium phosphate solutions. *Trans. Faraday Soc.* **1963**, *59*, 1585–1589. [CrossRef]
145. Norenberg, M. Oxidative and nitrosative stress in ammonia neurotoxicity. *Hepatology* **2003**, *37*, 245–248. [CrossRef]
146. Felipo, V.; Butterworth, R.F. Mitochondrial dysfunction in acute hyperammonemia. *Neurochem. Int.* **2002**, *40*, 487–491. [CrossRef]
147. Niknahad, H.; Jamshidzadeh, A.; Heidari, R.; Zarei, M.; Ommati, M.M. Ammonia-induced mitochondrial dysfunction and energy metabolism disturbances in isolated brain and liver mitochondria, and the effect of taurine administration: Relevance to hepatic encephalopathy treatment. *Clin. Exp. Hepatol.* **2017**, *3*, 141–151. [CrossRef] [PubMed]
148. Lai, J.C.K; Cooper, A.J.L. Neurotoxicity of ammonia and fatty acids: Differential inhibition of mitochondrial dehydrogenases by ammonia and fatty acyl coenzyme a derivatives. *Neurochem. Res.* **1991**, *16*, 795–803. [CrossRef] [PubMed]
149. Hird, F.J.R.; Marginson, M.A. Formation of Ammonia from Glutamate by Mitochondria. *Nature* **1964**, *201*, 1224–1225. [CrossRef] [PubMed]
150. Beavon, J.; Heatley, N.G. The occurrence of struvite (magnesium ammonium phosphate hexahydrate) in microbial cultures. *J. Gen. Microbiol.* **1962**, *31*, 167–169. [CrossRef] [PubMed]
151. Rivadeneyra, M.A.; Ramos-Cormenzana, A.; García-Cervigón, A. Bacterial formation of struvite. *Geomicrobiol. J.* **1983**, *3*, 151–163. [CrossRef]
152. Holt, L.E.; Pierce, J.A.; Kajdi, C.N. The solubility of the phosphates of strontium, barium, and magnesium and their relation to the problem of calcification. *J. Colloid Sci.* **1954**, *9*, 409–426. [CrossRef]
153. Racz, G.J.; Soper, R.J. Solubility of dimagnesium phosphate trihydrate and dimagnesium phosphate. *Can. J. Soil Sci.* **1968**, *48*, 265–269. [CrossRef]
154. Tabor, H.; Hastings, A.B. The ionization constant of secondary magnesium phosphate. *Biol. Chem.* **1943**, *148*, 627–632.
155. Verbeeck, R.M.H.; De Bruyne, P.A.M.; Driessens, F.C.M.; Verbeek, F. Solubility of magnesium hydrogen phosphate trihydrate and ion-pair formation in the system magnesium hydroxide-phosphoric acid-water at 25 °C. *Inorg. Chem.* **1984**, *23*, 1922–1926. [CrossRef]
156. Taylor, A.W.; Frazier, A.W.; Gurney, E.L. Solubility products of magnesium ammonium and magnesium potassium phosphates. *Trans. Faraday Soc.* **1963**, *59*, 1580–1584. [CrossRef]

157. Crosby, C.H.; Bailey, J. The role of microbes in the formation of modern and ancient phosphatic mineral deposits. *Front. Microbiol.* **2012**, *3*, e241–e247. [CrossRef] [PubMed]
158. Van Wazer, J.R.; Campanella, D.A. Structure and properties of the condensed phosphates. IV. Complex ion formation in polyphosphate solutions. *J. Am. Chem. Soc.* **1950**, *72*, 655–663. [CrossRef]
159. Kulakovskaya, T.V.; Lichko, L.P.; Vagabov, V.M.; Kulaev, I.S. Inorganic polyphosphates in mitochondria. *Biochemistry (Moscow)* **2010**, *75*, 825–831. [CrossRef] [PubMed]
160. Kornberg, A.; Rao, N.N.; Ault-Riché, D. Inorganic Polyphosphate: A Molecule of Many Functions. *Ann. Rev. Biochem.* **1999**, *68*, 89–125. [CrossRef] [PubMed]
161. Beauvoit, B.; Rigoulet, M.; Guerin, B.; Canioni, P. Polyphosphates as a source of high energy phosphates in yeast mitochondria: A^{31}P NMR study. *FEBS Lett.* **1989**, *252*, 17–21. [CrossRef]
162. Mansurova, S. E. Inorganic pyrophosphate in mitochondrial metabolism. *Biochim. Biophys. Acta (BBA) Bioenerg.* **1989**, *977*, 237–247. [CrossRef]
163. Pavlov, E.; Aschar-Sobbi, R.; Campanella, M.; Turner, R.J.; Gómez-García, M.R.; Abramov, A.Y. Inorganic Polyphosphate and Energy Metabolism in Mammalian Cells. *J. Biol. Chem.* **2010**, *285*, 9420–9428. [CrossRef]
164. Bobtelsky, M.; Kertes, S. The polyphosphates of calcium, strontium, barium and magnesium: Their complex character, composition and behaviour. *J. Appl. Chem.* **1954**, *4*, 419–429. [CrossRef]
165. Omelon, S.; Grynpas, M. Polyphosphates affect biological apatite nucleation. *Cells Tissues Organs* **2011**, *194*, 171–175. [CrossRef]
166. Docampo, R.; de Souza, W.; Miranda, K.; Rohloff, P.; Moreno, S.N.J. Acidocalcisomes–conserved from bacteria to man. *Nat. Rev. Microbiol.* **2005**, *3*, 251–261. [CrossRef] [PubMed]
167. Fleisch, H.; Neuman, W.F. Mechanisms of calcification: Role of collagen, polyphosphates, and phosphatase. *Am. J. Physiol.* **1961**, *200*, 1296–1300. [CrossRef] [PubMed]
168. Francis, M. The inhibition of calcium hydroxyapatite crystal growth by polyphosphonates and polyphosphates. *Calcif. Tissue Res.* **1969**, *3*, 151–162. [CrossRef] [PubMed]
169. Bunker, B.C.; Arnold, G.W.; Wilder, J.A. Phosphate glass dissolution in aqueous solutions. *J. Non-Cryst. Solids* **1984**, *64*, 291–316. [CrossRef]
170. Omelon, S.; Ariganello, M.; Bonucci, E.; Grynpas, M.; Nanci, A. A Review of Phosphate Mineral Nucleation in Biology and Geobiology. *Calcif. Tissue Int.* **2013**, *93*, 382–396. [CrossRef]
171. Reusch, R. N. Poly- -hydroxybutyrate/Calcium Polyphosphate Complexes in Eukaryotic Membranes. *Exp. Biol. Med.* **1989**, *191*, 377–381. [CrossRef]
172. Pavlov, E.; Zakharian, E.; Bladen, C.; Diao, C.T.M.; Grimbly, C.; Reusch, R.N.; French, R.J. A large, voltage-dependent channel, isolated from mitochondria by water-free chloroform extraction. *Biophys. J.* **2005**, *88*, 2614–2625. [CrossRef]
173. Reusch, R. N. Transmembrane Ion Transport by Polyphosphate/Poly-(R)-3-hydroxybutyrate Complexes. *Biochemistry (Moscow)* **2000**, *65*, 280–295.
174. Vagabov, V.M.; Trilisenko, L.V.; Kulaev, I.S. Dependence of inorganic polyphosphate chain length on the orthophosphate content in the culture medium of the yeast *Saccharomyces cerevisae*. *Biochemistry* **2000**, *65*, 349–354.
175. Nancollas, G.H. The mechanism of precipitation of biological minerals. The phosphates, oxalates and carbonates of calcium. *Croat. Chim. Acta* **1983**, *56*, 741–752.
176. Brečević, L.; Füredi-Milhofer, H. Precipitation of calcium phosphates from electrolyte solutions. *Calcif. Tissue Res.* **1972**, *10*, 82–90. [CrossRef] [PubMed]
177. Söhnel, O.; Mullin, J.W. Precipitation of calcium carbonate. *J. Crystal Growth* **1982**, *60*, 239–250. [CrossRef]
178. Rodriguez-Blanco, J.D.; Shaw, S.; Bots, P.; Roncal-Herrero, T.; Benning, L.G. The role of Mg in the crystallization of monohydrocalcite. *Geochim. Cosmochim. Acta* **2014**, *127*, 204–220. [CrossRef]
179. Kitamura, M. Crystallization and Transformation Mechanism of Calcium Carbonate Polymorphs and the Effect of Magnesium Ion. *J. Colloid Interface Sci.* **2001**, *236*, 318–327. [CrossRef] [PubMed]
180. Bots, P.; Benning, L.G.; Rodriguez-Blanco, J.D.; Roncal-Herrero, T.; Shaw, W. Mechanistic Insights into the Crystallization of Amorphous Calcium Carbonate (ACC). *Cryst. Growth Des.* **2012**, *12*, 3806–3814. [CrossRef]
181. Rodriguez-Blanco, J.D.; Shaw, S.; Bots, P.; Roncal-Herrero, T.; Benning, L.G. The role of pH and Mg on the stability and crystallization of amorphous calcium carbonate. *J. Alloys Compd.* **2012**, *536*, S477–S479. [CrossRef]

182. Nishiyama, R.; Munemoto, T.; Fukushi, K. Formation condition of monohydrocalcite from $CaCl_2$–$MgCl_2$–Na_2CO_3 solutions. *Geochim. Cosmochim. Acta* **2013**, *100*, 217–231. [CrossRef]
183. Rodriguez-Blanco, J.D.; Shaw, S.; Benning, L.G. The realtime kinetics and mechanisms of nucleation and growth of dolomite from solution. *Geochim. Cosmochim. Acta* **2009**, *73*, A1111.
184. Bischoff, J.L.; Fitzpatrick, J.A.; Rosenbauer, R.J. The solubility and stabilization of ikaite ($CaCO_3.6H_2O$) from 0° to 25 °C. *J. Geol.* **1992**, *101*, 21–33. [CrossRef]
185. Njegić-Džakula, B.; Brečević, L.; Falini, G.; Kralj, D. Kinetic Approach to Biomineralization: Interactions of Synthetic Polypeptides with Calcium Carbonate Polymorphs. *Croatica Chemica Acta* **2011**, *84*, 301–314. [CrossRef]
186. Cailleau, P.; Dragone, D.; Girou, A.; Humbert, L.; Jacquin, C.; Roques, H. Etude expérimentale de la précipitation des carbonates de calcium en présence de l'ion magnesium. *Bull. Soc. Fran. Mineral. Crystallogr.* **1977**, *100*, 81–88. [CrossRef]
187. Rivadeneyra, M.A.; Delgado, R.; Quesada, E.; Ramos-Cormenzana, A. Precipitation of calcium carbonate by Deleya halophila in media containing NaCl as sole salt. *Curr. Microbiol.* **1991**, *22*, 185–190. [CrossRef]
188. White, T.J.; Dong, Z.L. Structural derivation and crystal chemistry of apatite. *Acta Cryst. B* **2003**, *59*, 1–16. [CrossRef] [PubMed]
189. McArthur, J.M. Francolite geochemistry—compositional controls during formation, diagenesis, metamorphism and weathering. *Geochim. Cosmochim. Acta* **1985**, *49*, 23–35. [CrossRef]
190. Benmore, R.A.; Coleman, M.L.; McArthur, J.M. Origin of is sedimentary francolite from its sulphur and carbon isotope composition. *Nature* **1983**, *302*, 516–518. [CrossRef]
191. Ferguson, J.F.; McCarty, P.L. Effects of carbonate and magnesium on calcium phosphate precipitation. *Environ. Sci. Technol.* **1971**, *5*, 534–540. [CrossRef]
192. Van der Houwen, J.A.M.; Cressey, G.; Cressey, B.A.; Valsami-Jones, E. The effect of organic ligands on the crystallinity of calcium phosphate. *J. Cryst. Growth* **2003**, *249*, 572–583. [CrossRef]
193. Suchanek, W.J.; Byrappa, K.; Shuk, P.; Riman, R.E.; Janas, V.F.; Tenhuisen, K.S. Mechanochemical-hydrothermal synthesis of calcium phosphate powders with coupled magnesium and carbonate substitution. *J. Solid State Chem.* **2004**, *17*, 793–799. [CrossRef]
194. Fleet, M.E.; Liu, X. Coupled substitution of type A and B carbonate in sodium-bearing apatite. *Biomaterials* **2007**, *28*, 916–926. [CrossRef]
195. LeGeros, R.Z.; Trautz, O.R.; Klein, E.; LeGeros, J.P. Two types of carbonate substitution in the apatite structure. *Experientia* **1969**, *25*, 5–7. [CrossRef]
196. Rey, C.; Collins, B.; Goehl, T.; Dickson, R.I.; Glimcher, M.J. The carbonate environment in bone mineral: A resolution-enhanced fourier transform infrared spectroscopy study. *Calcif. Tissue Int.* **1989**, *45*, 157–164. [CrossRef] [PubMed]
197. Trautz, O.R. Crystallographic studies of calcium carbonate phosphate. *Ann. N. Y. Acad. Sci.* **1960**, *85*, 145–160. [CrossRef] [PubMed]
198. Cao, X.; Harris, W.G.; Josan, M.S.; Nair, V.D. Inhibition of calcium phosphate precipitation under environmentally-relevant conditions. *Sci. Total Environ.* **2007**, *383*, 205–215. [CrossRef]
199. Legros, R.; Balmain, N.; Bonel, G. Age-related changes in mineral of rat and bovine cortical bone. *Calcif. Tissue Int.* **1987**, *41*, 137–144. [CrossRef]
200. Wu, Y.; Glimcher, M.J.; Rey, C.; Ackerman, J.L. A unique protonated phosphate group in bone mineral not present in synthetic calcium phosphates. Identification by phosphorus-31 solid state NMR spectroscopy. *J. Mol. Biol.* **1994**, *244*, 423–435. [CrossRef]
201. Rey, C.; Shimizu, M.; Collins, B.; Glimcher, M.J. Resolution-enhanced Fourier transform infrared spectroscopy study of the environment of phosphate ions in the early deposits of a solid phase of calcium phosphate in bone and enamel, and their evolution with age. I: Investigations in the ν_4 PO_4 domain. *Calcif. Tissue Int.* **1990**, *46*, 384–394. [CrossRef]
202. Montel, G.; Bonel, G.; Heughebaert, J.C.; Trombe, J.C.; Rey, C. New concepts in the composition, crystallization and growth of the mineral component of calcified tissues. *J. Cryst. Growth* **1981**, *53*, 74–99. [CrossRef]
203. Rey, C.; Combes, C.; Drouet, C.; Sfihi, H. Chemical Diversity of Apatites. *Adv. Sci. Tech.* **2006**, *49*, 27–36. [CrossRef]
204. Combes, C.; Cazalbou, S.; Rey, C. Apatite Biominerals. *Minerals* **2016**, *6*, 34. [CrossRef]

205. Cazalbou, S.; Eichert, D.; Ranz, X.; Drouet, C.; Combes, C.; Harmand, M.F.; Rey, C. Ion exchanges in apatites for biomedical application. *J. Mater. Sci. Mater. Med.* **2005**, *16*, 405–409. [CrossRef]
206. Eichert, D.; Combes, C.; Drouet, C.; Rey, C. Formation and evolution of hydrated surface layers of apatites. *Key Eng. Mater.* **2005**, *284–286*, 3–6. [CrossRef]
207. Busenberg, E.; Plummer, L.N. Thermodynamics of magnesian calcite solid-solutions at 25 °C and 1 atm total pressure. *Geochim. Cosmochim. Acta* **1989**, *53*, 1189–1208. [CrossRef]
208. Bischoff, W.D.; Mackenzie, F.T.; Bishop, F.C. Stabilities of synthetic magnesian calcites in aqueous solution: Comparison with biogenic materials. *Geochim. Cosmochim. Acta* **1987**, *51*, 1413–1423. [CrossRef]
209. Bertram, M.A.; Mackenzie, F.T.; Bishop, F.C.; Bischoff, W.D. Influence of temperature on the stability of magnesian calcite. *Am. Mineral.* **1991**, *76*, 1889–1896.
210. Tribble, J.S.; Arvidson, R.S.; Lane, M.; Mackenzie, F.T. Crystal chemistry, and thermodynamic and kinetic properties of calcite, dolomite, apatite, and biogenic silica: Applications to petrologic problems. *Sediment. Geol.* **1995**, *95*, 11–37. [CrossRef]
211. Michałowski, T.; Asuero, A.G. Thermodynamic Modelling of Dolomite Behavior in Aqueous Media. *J. Thermodyn.* **2012**, *2012*, 1–12. [CrossRef]
212. Moomaw, A.S.; Maguire, M.E. The unique nature of Mg^{2+} channels. *Physiology (Bethesda)* **2008**, *23*, 275–285. [CrossRef]
213. Usdovski, E. Synthesis of dolomite and geochemical implications. In *Dolomites: A Volume in Honour of Dolomieu*; Purser, B.H., Tucker, M.E., Zenger, D.H., Eds.; Blackwell Sci. Publ.: Oxford, UK, 1994; pp. 345–360.
214. Sherman, L.A.; Barak, P. Solubility and dissolution kinetics of dolomite in $Ca-Mg-HCO_3/CO_3$ solutions at 25 °C and 0.1 MPa carbon dioxide. *Soil Sci. Soc. Am. J.* **2000**, *64*, 1959–1968. [CrossRef]
215. Pabst, A. The Crystallography Structure of Eitelite, $Na_2Mg(CO_3)_2$. *Am. Miner.* **1973**, *58*, 211–217.
216. Effenberger, H. Crystal-structure and infrared-absorption spectrum of synthetic monohydrocalcite, $CaCO_3 \cdot H_2O$. *Monatsh. Chem.* **1981**, *112*, 899–909. [CrossRef]
217. Swainson, I. P. The structure of monohydrocalcite and the phase composition of the beachrock deposits of Lake Butler and Lake Fellmongery, South Australia. *Am. Mineral.* **2008**, *93*, 1014–1018. [CrossRef]
218. Bachra, B.N.; Trautz, O.R.; Simon, S.L. Precipitation of calcium carbonates and phosphates. I. Spontaneous precipitation of calcium carbonates and phosphates under physiological conditions. *Arch. Biochem. Biophys.* **1963**, *103*, 124–138. [CrossRef]
219. Ferguson, J.F.; McCarty, P.L. *The Precipitation of PhosDhates from Fresh Waters and Waste Waters*; Tech Report No. 120; Department of Civil Engineering, Stanford University: Stanford, CA, USA, 1969.
220. Kibalczyc, W.; Christoffersen, J.; Christoffersen, M.R.; Zielenkiewicz, A.; Zielenkiewicz, W. The effect of magnesium-ions on the precipitation of calcium phosphates. *J. Cryst. Growth* **1990**, *106*, 355–366. [CrossRef]
221. Bachra, B.N.; Trautz, O.R.; Simon, S.L. Precipitation of calcium carbonates and phosphates—III. *Arch. Oral Biol.* **1965**, *10*, 731–738. [CrossRef]
222. Mekmene, O.; Quillard, S.; Rouillon, T.; Bouler, J.-M.; Piot, M.; Gaucheron, F. Effects of pH and Ca/P molar ratio on the quantity and crystalline structure of calcium phosphates obtained from aqueous solutions. *Dairy Sci. Technol. EDP Sci./Springer* **2009**, *89*, 301–316. [CrossRef]
223. Terpstra, R.A.; Driessens, F.C.M. Magnesium in tooth enamel and synthetic apatites. *Calcif. Tissue Int.* **1986**, *39*, 348–354. [CrossRef]
224. Martens, C.S.; Harriss, R.C. Inhibition of apatite precipitation in the marine environment by magnesium ions. *Geochim. Cosmochim. Acta* **1970**, *34*, 621–625. [CrossRef]
225. Termine, J.D.; Conn, K.M. Inhibition of apatite formation by phosphorylated metabolites and macromolecules. *Calcif. Tiss. Res.* **1976**, *22*, 149–157. [CrossRef]
226. Posner, A.S.; Betts, F.; Blumenthal, N.C. Role of ATP and Mg in the stabilization of biological and synthetic amorphous calcium phosphates. *Calcif. Tissue Res.* **1977**, *22*, 208–212. [CrossRef]
227. Rey, C.; Trombe, J.C.; Montel, G. Some features of the incorporation of oxygen in different oxidation states in the apatitic lattice—III Synthesis and properties of some oxygenated apatites. *J. Inorg. Nucl. Chem.* **1978**, *40*, 27–30. [CrossRef]
228. Pan, Y.; Fleet, M.E. Compositions of the Apatite-Group Minerals: Substitution Mechanisms and Controlling Factors. *Rev. Mineral. Geochem.* **2002**, *48*, 13–49. [CrossRef]

229. Mason, A.Z.; Nott, J.A. The role of intracellular biomineralized granules in the regulation and detoxification of metals in gastropods with special reference to the marine prosobranch *Littorina littorea*. *Aquat. Toxicol.* **1981**, *1*, 239–256. [CrossRef]
230. Nott, J.A.; Langston, W.J. Cadmium and the phosphate granules in *Littorina littorea*. *J. Mar. Biol. Assoc. UK* **1989**, *69*, 219–227. [CrossRef]
231. Bonucci, E.; Derenzini, M.; Marinozzi, V. The organic-inorganic relationship in calcified mitochondria. *J. Cell. Biol.* **1973**, *59*, 185–211. [CrossRef]
232. Baartscheer, A.; Schumacher, C.A.; Coronel, R.; Fiolet, J.W.T. The driving force of the Na^+/Ca^{2+}-exchanger during metabolic inhibition. *Front. Physiol.* **2011**, *2*, 1–7. [CrossRef] [PubMed]
233. Griffiths, E. J.; Halestrap, A. P. Mitochondrial non-specific pores remain closed during cardiac ischaemia, but open upon reperfusion. *Biochem. J.* **1995**, *307*, 93–98. [CrossRef]
234. Bernardi, P. Mitochondrial transport of cations: Channels, exchangers, and permeability transition. *Physiol. Rev.* **1999**, *79*, 1127–1155. [CrossRef]
235. Meldrum, D.R.; Cleveland, J.C.J.; Mitchell, M.B.; Sheridan, B.C.; Gamboni-Robertson, F.; Harken, A.H.; Banerjee, A. Protein kinase C mediates Ca^{2+}-induced cardioadaptation to ischemia-reperfusion injury. *Am. J. Physiol.* **1996**, *271*, R718–R726. [CrossRef]
236. Miyawaki, H.; Zhou, X.; Ashraf, M. Calcium preconditioning elicits strong protection against ischemic injury via protein kinase C signaling pathway. *Circ. Res.* **1996**, *79*, 137–146. [CrossRef]

© 2020 by the authors. Licensee MDPI, Basel, Switzerland. This article is an open access article distributed under the terms and conditions of the Creative Commons Attribution (CC BY) license (http://creativecommons.org/licenses/by/4.0/).

Review

Solid-Contact Ion-Selective Electrodes: Response Mechanisms, Transducer Materials and Wearable Sensors

Yan Lyu [1], Shiyu Gan [1,*], Yu Bao [1], Lijie Zhong [1], Jianan Xu [2,3], Wei Wang [1], Zhenbang Liu [1], Yingming Ma [1], Guifu Yang [4] and Li Niu [1,5,*]

1. School of Civil Engineering, c/o Center for Advanced Analytical Science,
   School of Chemistry and Chemical Engineering, Guangzhou University, Guangzhou 510006, China;
   yanlyu@e.gzhu.edu.cn (Y.L.); baoyu@gzhu.edu.cn (Y.B.); ccljzhong@gzhu.edu.cn (L.Z.);
   wangw@gzhu.edu.cn (W.W.); cczbliu@gzhu.edu.cn (Z.L.); ccymma@gzhu.edu.cn (Y.M.)
2. State Key Laboratory of Electroanalytical Chemistry, c/o Engineering Laboratory for Modern Analytical Techniques, Changchun Institute of Applied Chemistry, Chinese Academy of Sciences, Changchun 130022, China; jnxu@ciac.ac.cn
3. University of Chinese Academy of Sciences, Beijing 100039, China
4. School of Information Science and Technology, Northeast Normal University, Changchun 130117, China; young@nenu.edu.cn
5. MOE Key Laboratory for Water Quality and Conservation of the Pearl River Delta, Guangzhou University, Guangzhou 510006, China
* Correspondence: ccsygan@gzhu.edu.cn (S.G.); lniu@gzhu.edu.cn (L.N.)

Received: 28 May 2020; Accepted: 16 June 2020; Published: 23 June 2020

Abstract: Wearable sensors based on solid-contact ion-selective electrodes (SC-ISEs) are currently attracting intensive attention in monitoring human health conditions through real-time and non-invasive analysis of ions in biological fluids. SC-ISEs have gone through a revolution with improvements in potential stability and reproducibility. The introduction of new transducing materials, the understanding of theoretical potentiometric responses, and wearable applications greatly facilitate SC-ISEs. We review recent advances in SC-ISEs including the response mechanism (redox capacitance and electric-double-layer capacitance mechanisms) and crucial solid transducer materials (conducting polymers, carbon and other nanomaterials) and applications in wearable sensors. At the end of the review we illustrate the existing challenges and prospects for future SC-ISEs. We expect this review to provide readers with a general picture of SC-ISEs and appeal to further establishing protocols for evaluating SC-ISEs and accelerating commercial wearable sensors for clinical diagnosis and family practice.

Keywords: ion selective electrodes; wearable sensors; solid-contact materials; response mechanism

---

1. Introduction

With the rapid growth of personal healthcare and fitness systems, wearable devices that can provide real-time and continuous monitoring of an individual's physiological state have attracted great attention in recent years. Conventional applications in medical biomarker detection generally related to separated collection and analysis of blood samples is invasive and costly and needs complicated operations, while failing to provide users with real-time health diagnostics and monitoring. Wearable sensors hold great promise for continuously monitoring an individual's physiological bio-chemical signals [1–5].

A variety of flexible sensors have recently been developed for non-invasively assessing personal physiological states by detecting analytes of interest (like those in sweat) [6–9]. Potentiometric sensors, particularly for the ion sensing, are one of the attractive types for practical application due to their high portability, good sensitivity, lower energy consumption and high efficiency. Potentiometry incorporates a working electrode (WE) and a reference electrode (RE) and measures their relative potential under zero current. Ion-selective electrodes (ISEs) are the typical potentiometric sensor for selective ion recognition.

Classic liquid-contact ISEs (LC-ISEs) contain an ion-selective membrane (ISM, e.g., pH glass membrane) and an internal solution (Figure 1A) to form a liquid–contact interface [10]. The theoretical fundamental of ISEs is based on the relationship between ion activity and output voltage according to the Nernst equation. The electromotive force (EMF) is the sum of all the phase boundary potentials. LC-ISEs have been commercialized and quite popular for various ion analysis in the laboratory and environmental analysis. However, the biggest challenge for the LC-ISEs is miniaturization and integration, which could not satisfy the requirements in biological relative applications, like cell or tissue-level ion analysis and wearable sensors. Cattrall and Freiser fabricated the first solid-contact ISEs (SC-ISEs) without the internal solution in 1971, which they called "coated wire electrodes" (CWEs) [11]. A quite simple electrode structure was proposed, i.e., a metal wire directly coated with Ca^{2+} ionophore-containing polymeric sensing membrane. The CWEs exhibited the exciting Nernstian response toward Ca^{2+}. Freiser et al. further extended the CWEs for other ions [12]. Although the large potential drift is attributed to the unstable potential at the metal/membrane interface, the milestone CWEs opened the way for the SC-ISEs [13–15].

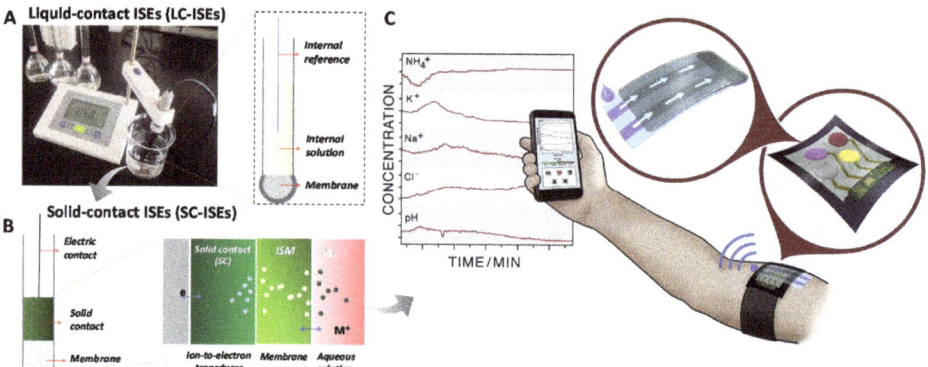

Figure 1. An overview from liquid-contact ion-selective electrodes (LC-ISEs) to solid-contact ISEs (SC-ISEs) for wearable sensors. (**A**) Classic LC-ISEs (e.g., pH meter) by a liquid-contact between internal solution and ion-selective membrane (ISM). (**B**) The structure of SC-ISEs by a solid-contact between solid ion-to-electron transducer layer and ISM. (**C**) An example of the SC-ISEs for wearable sensor applications [9].

In 1992, Lewenstam and Ivaska et al. further focused on this issue and proposed an intermediate polypyrrole (PPy) [16] solid-contact layer between an ISM and conducting substrate, called the "ion-to-electron transducer layer" (Figure 1B). This transducer can transfer the ion concentration to electron signal and stabilizes the potential at the substrate/ISM interface. This electrode structure has become the-state-of-the-art standard model of SC-ISEs. Currently, numerous materials have been developed for the transducer, such as conducting polymers, multifarious carbon materials, nanomaterials, molecular redox couples etc. The potential stability has been remarkably improved and the applications for SC-ISEs are also attracting increasing attention. A representative application is for wearable sensors [9] (Figure 1C). Since the all-solid-state structure is the characteristic for the SC-ISEs, this feature makes the SC-ISEs flexible, stretchable and miniaturized.

This review concentrates on the recent advances of SC-ISEs for wearable sensors in real-time and non-invasive analysis of ions in biological fluids. Before that, we will illustrate the response mechanisms and the crucial solid-contact transducer materials. In the response mechanism section, the two mechanisms including redox capacitance and electric-double-layer (EDL) capacitance-based potential stabilization will be discussed. For the solid-contact transducer materials, classic conductive polymers and carbon materials and other nanomaterials (e.g., Au nanomaterials) will be illustrated in detail. In the wearable sensor sections, we will mainly discuss the flexible SC-ISEs for real-time monitoring ion in biological fluids, mainly sweat ion analysis. Finally, we summarize the current developments, existing challenges, and future prospects for SC-ISEs-based portable and wearable sensors.

2. Response Mechanisms

Like the conventional LC-ISEs, the response mechanism for the SC-ISEs is the sum of the phase interfacial potentials. As shown in Figure 2, there are two typical response mechanisms [14] for the ion-to-electron single transformation including the redox capacitance mechanism (Figure 2A) and the EDL capacitance mechanism (Figure 2B). These two mechanisms have been verified experimentally [17,18]. Herein we briefly illustrate the mechanisms through the phase interfacial potential. The response mechanism generally depends on the SC materials.

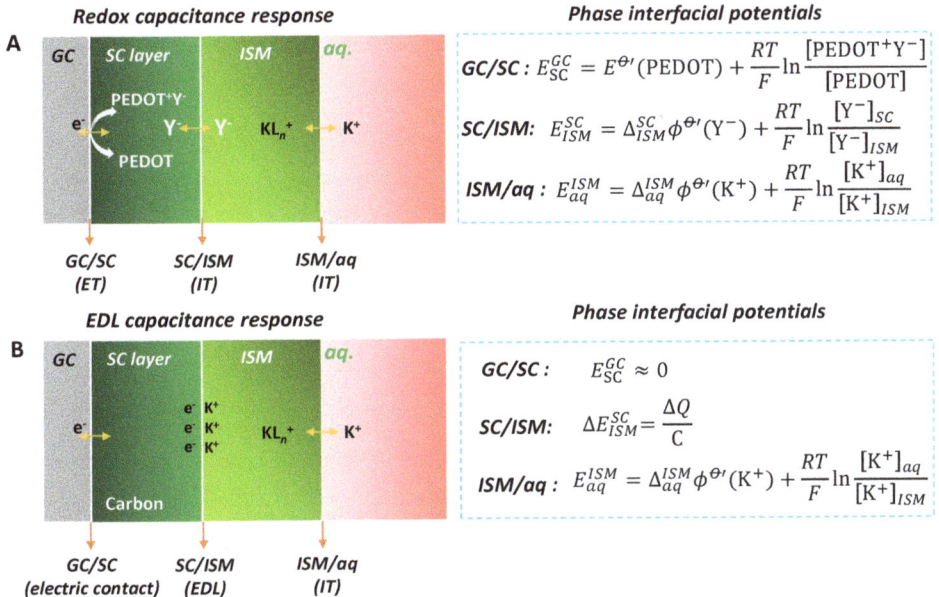

Figure 2. Response mechanisms for the SC-ISEs. (**A**) Redox capacitance-based SC-ISEs with poly(3,4-ethylenedioxythiophene) (PEDOT) as an example for redox SC transducer. (**B**) Electric-double-layer (EDL) capacitance-based SC-ISEs with carbon as an example for EDL SC transducer. Both SC-ISEs contain three interfaces, GC/SC, SC/ISM and ISM/aq. GC: glass carbon electrode substrate; SC: solid contact; aq: aqueous solution; ET: electron transfer; IT: ion transfer. The corresponding phase interfacial potentials are presented on the right (detailed illustration shown in the main text and Appendix A).

Redox capacitance mechanism. For the redox capacitance-based SC-ISEs, the SC materials disclose highly reversible redox behavior and can effectively attach on the surface of the substrate electrode, for example, the typical poly(3,4-ethylenedioxythiophene) (PEDOT) redox conductive polymer. These materials exhibit both electric and ionic conductivities. They convert the target ion concentration to an electron signal through the oxidation or reduction reaction. Taking the Y^-

anion-doped PEDOT and K⁺ response as an example (Figure 2A), the ion-to-electron response process can be described as follows:

$$\text{PEDOT}^+\text{Y}^-(\text{SC}) + \text{K}^+(aq) + e^-(\text{GC}) \rightleftharpoons \text{PEDOT}(\text{SC}) + \text{Y}^-(\text{ISM}) + \text{K}^+(\text{ISM}) \tag{1}$$

where PEDOT⁺Y⁻ (SC) and PEDOT (SC) represent the oxidation and reduced states of PEDOT in the SC phase, respectively; K⁺ (aq) and K⁺ (ISM) are the concentrations in the measured aqueous solution and ISM phase, respectively; Y⁻ (SC) and e⁻ (glass carbon, GC) present the Y⁻ in the SC and GC electrode phases, respectively.

It should be noted that the above Equation (1) represents the overall ion-to-electron reaction. In fact, the overall reaction involves three equilibrium charge transfers at three interfaces (or phase boundary). The first is the electron transfer (ET) of PEDOT at the GC/SC interface. The GC/SC interfacial potential E_{SC}^{GC} can be described according to the Nernst equation,

$$E_{SC}^{GC} = E^{O\prime}(\text{PEDOT}) + \frac{RT}{F} \ln \frac{[\text{PEDOT}^+\text{Y}^-]}{[\text{PEDOT}]} \tag{2}$$

where $E^{O\prime}(\text{PEDOT})$ is the conditional electrode potential of PEDOT; R, T and F represent the gas constant, temperature and Faradaic constant, respectively. Once the PEDOT is attached on the GC electrode, their concentrations are fixed, leading to a constant potential of E_{SC}^{GC}.

The second equilibrium charge transfer is the ion transfer (IT) of doped Y⁻ anion at the SC/ISM interface. For example, taking the tetrakis-(pentafluorophenyl)borate (TPFPhB⁻) anion-doped PEDOT as an example, the TPFPhB⁻ is the doped counter ion of PEDOT in the solid contact phase and also the electrolyte of KTPFPhB in the ISM phase. The TPFPhB⁻ anion in the two phases reaches the IT equilibrium state. The SC/ISM interfacial potential E_{ISM}^{SC} can be described as follows (the detailed derivation, see Appendix A),

$$E_{ISM}^{SC} = \Delta_{ISM}^{SC}\phi^{O\prime}(\text{Y}^-) + \frac{RT}{F} \ln \frac{[\text{Y}^-]_{SC}}{[\text{Y}^-]_{ISM}} \tag{3}$$

where $\Delta_{ISM}^{SC}\phi^{O\prime}(\text{Y}^-)$ is the conditional ion transfer potential of Y⁻ anion from SC to ISM phase; $[\text{Y}^-]_{ISM}$ and $[\text{Y}^-]_{SC}$ represent the Y⁻ concentrations in the ISM and SC phases, respectively. It should be noted that multi-ion transfer equilibrium could exist at the SC/ISM interface. If that happens, multi-ion transfer Nernst equations like Equation (3) should be presented. The SC/ISM interfacial potential is determined by the sum of multi-ion transfer Nernst equations. Once the SC and ISM are attached, the Y⁻ concentration are fixed and the E_{ISM}^{SC} is also a constant.

The third equilibrium charge transfer is the IT of K⁺ at the ISM/aq interface. Similar to Y⁻ at SC/ISM interface, the ISM/aq interfacial potential is described as follows,

$$E_{aq}^{ISM} = \Delta_{aq}^{ISM}\phi^{O\prime}(\text{K}^+) + \frac{RT}{F} \ln \frac{[\text{K}^+]_{aq}}{[\text{K}^+]_{ISM}} \tag{4}$$

where $\Delta_{aq}^{ISM}\phi^{O\prime}(\text{K}^+)$ is the standard ion transfer potential of K⁺ from ISM to aqueous phase; $[\text{K}^+]_{aq}$ and $[\text{K}^+]_{ISM}$ represent the K⁺ concentration in the aqueous and ISM phases, respectively. It should be noted that the IT of K⁺ dominates the ISM/aq interfacial potential. There could not be multi-ion transfer equilibrium at this interface since the ionophore can selectively recognize the K⁺.

The potential for the SC-ISEs (E) is the sum of all the three interfacial potentials (Equations (2)–(4)), which is described as follows,

$$E = E_{SC}^{GC} + E_{ISM}^{SC} + E_{aq}^{ISM} = k + \frac{RT}{F} \ln[\text{K}^+]_{aq} \tag{5}$$

Once the SC and ISM components are fixed, except for the $\left[K^+\right]_{aq}$, the other items (Equations (2)–(4)) are the constant (k). Thus, the measured potential E shows the Nernstian response toward the target ion. We should here emphasize that the potential for redox capacitance-based SC-ISEs is thermodynamically defined according to Equation (5).

As the most popular SC transducer materials, a wide range of conducting polymers have been explored. In addition, other redox molecule and nanomaterials with effective ion-to-electron transduction have been developed, which will be discussed in Section 3.

EDL capacitance mechanism. For SC-ISEs with an EDL capacitance mechanism (Figure 2B), the ion-to-electron transduction is formed from the EDL between the ISM and the SC interface. It can be described as an asymmetric capacitor where one side is formed by electrons (or holes) and the other side is balanced by ions (cations or anions) from the ISM. The interfacial potential depends on the quantity of charge and EDL capacitance. Like redox capacitance-based SC-ISEs, there are also three interfaces including GC/SC, SC/ISM and ISM/aq interfaces (Figure 2B).

For the GC/SC interface, the potential is very small ($E_{SC}^{GC} \approx 0$) since most EDL capacitance-based SC materials are highly electronic conductive, for example carbon materials. For the SC/ISM interface, this interface has no charge transfer reaction so that this interfacial potential cannot be defined. It can only determine the potential variation, which is descried as follows,

$$\Delta E_{ISM}^{SC} = \frac{\Delta Q}{C} \tag{6}$$

where ΔQ is the passed charge and C is the EDL capacitance. For the ISM/aq interface, the potential is the same with Equation (3). Therefore, the potential stability for the EDL capacitance type SC-ISE depends on the SC/ISM interface. If the EDL capacitance is high, the potential change would be small or even approaches zero.

A straightforward approach to increase the EDL capacitance is increasing the interfacial contact area between the ISE membrane and the SC. For example, porous carbon materials and nanomaterials have been developed for increasing the EDL capacitance. We will discuss this in Section 3.

3. Transducer Materials

The development of SC transducer materials is the core subject of SC-ISEs. Conceptually, SC transducer works as an ion-to-electron signal transformation to provide operationally stable and reproducible potentials. For the wearable sensors, SC transducer materials require more properties, such as portability, excellent flexibility, and maintenance-free. The reversible and equilibrium stability on ion-to-electron transduction in the solid state requires high exchange currents compared with the current passed during the test. Numerous efforts have been devoted to exploring better SC materials with high conductivity, hydrophobicity, large capacitance, and high stability. Next, we will discuss the state-of-the-art SC transducer materials including conducting polymers (CPs), carbon-based materials and other typical nanomaterials.

3.1. Conducting Polymers

CPs are the most popular SC transducer materials for SC-ISEs. CPs were first introduced as efficient ion-to-electron transducers by Lewenstam and Ivaska groups in the early-1990s [16]. They proposed PPy as the SC. Nernstian response toward Na$^+$ sensing was resulted, and the stability lasted over 10 days. This work has enabled the establishment of CPs-based SC-ISEs with remarkably improved stability compared with the CWEs. Over the last three decades, CPs have been intensively explored for SC materials [13,19–23], such as PPy, PEDOT, poly(3-octylthiophene) (POT), polyaniline (PANI) and so on. One of the featured characteristics for CPs is their mixed electronic and ionic conductivities. The ionic signal can be effectively transformed to electronic signal through the redox of CPs coupled with ion transfer (Figure 2A). In addition, the stable potentiometric signals contribute to their high redox capacitance, which overcomes the small current passed through the electrode during zero-current conditions and environmental interruption. CPs could be easily electrodeposited on the electrode substrate and controlled by the chronocoulometry. The flexibility of CPs also becomes the advantage for the wearable sensors. However, there remain several challenges for the CP-based SC-ISEs: (1) undesired electrochemical side reactions leading to environmental sensitivity, such as gas or light; (2) the accumulation of a water layer at the SC/ISM interface.

To reduce the interfacial polarization of SC/ISM, Gyurcsányi and coworkers reported 3D nanostructured poly(3,4-ethylenedioxythiophene) polystyrene sulfonate (PEDOT(PSS)) film as a SC material for Ag^+ sensing by using 3D nanosphere lithography and electrosynthesis [24] (Figure 3A). The constructed 3D nanostructured PEDOT(PSS) film had almost 7 times larger redox capacitance than the one fabricated by direct electrosynthesis. In addition, the 3D PEDOT(PSS) was functionalized with a lipophilic redox couple (1,1′-dimethylferrocene) to further increase the redox capacitance and hydrophobicity. The EMF response revealed an improved standard deviation E^0 of 3.9 mV. Such a 3D structure offers high redox capacitance to enhance the reproducibility of SC-ISEs.

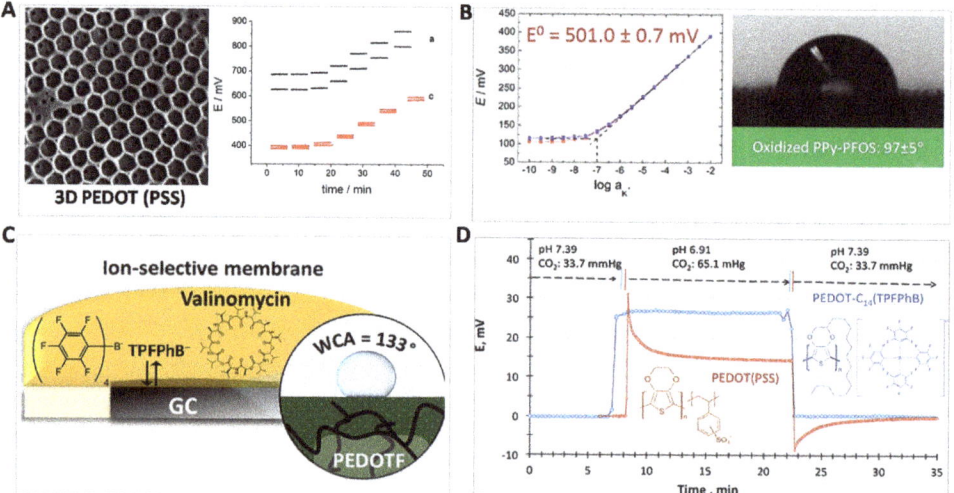

Figure 3. Structural design and functionalization of conducting polymers (CPs) for CP-based SC-ISEs. (**A**) Scanning electronic microscopy (SEM) image of the designed high-surface 3D poly(3,4-ethylenedioxythiophene) polystyrene sulfonate (PEDOT(PSS)) by nanosphere lithography and electrosynthesis. The right: the potential traces of 3D PEDOT(PSS) functionalized with (red line) and without (black line) a lipophilic redox 1,1′-dimethylferrocene. It is found that the functionalized 3D PEDOT(PSS) exhibits much better reproducibility. Reprinted with permission from [24], Copyright (2016) John Wiley and Sons publications. (**B**) The oxidized PPy by doping perfluorooctanesulfonate ($PFOS^-$) anion (PPy-PFOS) enhanced the hydrophobicity leading to remarkably improved reproducibility. Reprinted with permission from [25], Copyright (2017) American Chemical Society. (**C**) Superhydrophobic tetrakis-(pentafluorophenyl)borate ($TPFPhB^-$) anion doping PEDOT as an SC transducer to reduce the water-layer effect. It should be noted the $TPFPhB^-$ ion transfer (ion exchange) between SC and ISM further enhanced the potential stability (see Equation (3)). The abbreviation of tetrakis-(pentafluorophenyl)borate on the original Figure is $TFAB^-$. Herein it is replaced by $TPFPhB^-$. Reprinted with permission from [26], Copyright (2019) American Chemical Society. (**D**) An ultimate approach by both C_{14}-chain functionalized PEDOT and $TPFPhB^-$ doping to improve the performances of SC-ISEs. Reprinted with permission from [27], Copyright (2017) American Chemical Society.

The hydrophobicity for the classic CPs (like PEDOT and PPy) is in fact insufficient for eliminating the water layer. Tuning the counter ion doping is an efficient strategy to improve the intrinsic hydrophobicity of CPs. For example, Gyurcsányi and co-workers implemented a hydrophobic perfluorinated anion, perfluorooctanesulfonate (PFOS$^-$) as a doping ion in PPy, which increased the hydrophobicity [25] (contact angle: 97° ± 5°) (Figure 3B). The PFOS$^-$ ion doping resulted in an oxidized PPy that enhanced its conductivity. An improved reproducible standard potential E^0 of 501.0 ± 0.7 mV has been achieved, which is a very small standard deviation for CP-based SC-ISEs. Very recently, Lindfors and coworkers further introduced a more hydrophobic counter ion, tetrakis-(pentafluorophenyl)borate (TPFPhB$^-$) for doping PEDOT [26] (Figure 3C). The water contact angle is up to 133°. It should be noted that the TPFPhB$^-$ anion exists in both SC phase and ISM phase, which determines the SC/ISM interfacial potential (see Equation (3)). By virtue of TPFPhB$^-$ ion doping, the K$^+$ sensor showed a low drift of 50 µV/h over 49 days.

In addition to counter ion doping, Lindner and coworkers improved the framework of PEDOT and used TPFPhB$^-$ ion doping simultaneously [27] (Figure 3D). They functionalized PEDOT with a C_{14} alkane chain (PEDOT-C_{14}). The superhydrophobic counter ion, TPFPhB$^-$ was also doped into PEDOT-C_{14}. This strategy further improved the interfacial hydrophobicity (contact angel: 136° ± 5°). This SC material has been fabricated for pH, K$^+$ and Na$^+$ sensors. The long-term potential stability showed the lowest drift (0.02 ± 0.03 mV/day). For the pH response, it disclosed ±0.002 pH unit repeatability and more importantly, the CO_2 interference has been suppressed. We would comment that this work provides the ultimate approach to improving PEDOT-based SC materials.

Compared with relatively hydrophilic bulk CPs of PEDOT and PPy, POT is a highly hydrophobic CP. Ivaska et al. introduced POT [28] as the SC for Ca^{2+} sensing in 1994. The results showed a standard potential (259.3 ± 1.3 mV) and small drift (0.23 mV/day), which were comparable with conventional Ca-ISEs. Since POT is highly hydrophobic, the water-layer effect should be significantly suppressed [29]. However, the reported analytical performances are contradictions. For example, Lindner's group prepared POT-based K$^+$-SC-ISEs, displaying 1.4 mV/h drift [30]. After doping with a 7,7,8,8-tetracyanoquinodimethane (TCNQ/TCNQ$^-$) redox couple, the drift decreased to 0.1 mV/h. We speculate that it might attribute to the high resistance of POT, leading to a quasi-reversible (POT/POT^{n+}) particularly for the thicker POT. In addition, the light-sensitivity for POT compared with other CPs is the most challenging. It should be noted that Bakker's group has established voltammetry-based potentiometric ion sensing by using the ET of POT coupling with the IT of target ions [31–37], and first proposed POT-based transducer for both anion and cation sensing by electrochemical switching [38]. PANI has also been a widely used CP for SC-ISEs by virtue of its easy electrochemical synthesis and excellent electronic/ionic conductivities. A drawback is the pH-sensitivity. Liu et al. reported a composite PANI/PMMA (poly(methyl methacrylate)) to reduce this pH-effect [39]. Numerous efforts for functionalization of PANI composites [40–44] have been devoted to further enhancing the analytical performance. Other CPs, for example, polyazulene [45,46], ferrocene-functionalized poly(vinyl chloride) (PVC) [47] are also proposed for the SC-ISEs.

In addition to the SC materials, the leaking of ISM components is another daunting challenge for the SC-ISEs. The ISM is in fact a plasticized organic solid membrane. The membrane components face the risk of leaking into the solution. Michalska et al. experimentally observed the spontaneous partition of POT into the ISM by ultraviolet/visible (UV/vis) and fluorescence microscopy [48]. The POT was distributed into the whole ISM and the partition ratio reached 0.5% w/w. Lisak and coworkers recently reported an approach by coating an outer layer of inert silicone rubber (SR) on the ISM layer to reduce the leaking [49] (Figure 4). Their experimental results demonstrated that SR coating did not affect the selectivity and improved the analytical performances remarkably. The potential stability of standard potential increased from ±35.3 mV to 3.5 mV (Figure 4D). The Fourier transform infrared (FTIR) spectra demonstrated that partial ISM components gradually diffused into SR layer (Figure 4C). Since the lower solubility of ISM components in the SR, the leaking is reduced, leading to an enhanced stability. Moreover, the SR increased the resistance to biofouling which would be beneficial to the practical measurement in a complicated environment, for example in sea water.

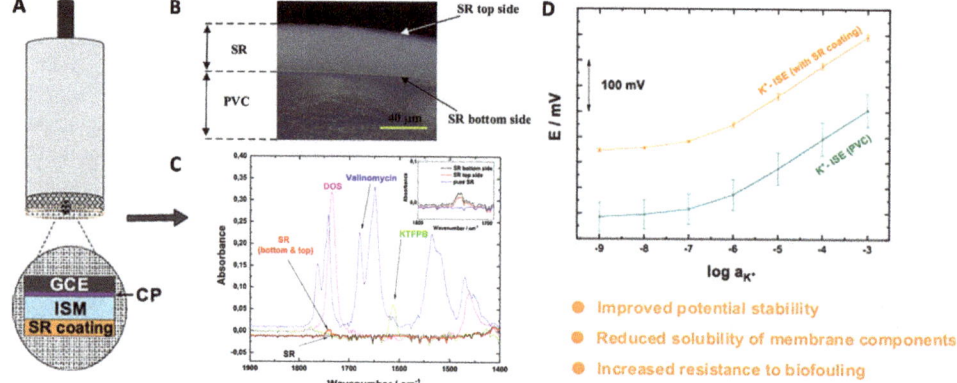

Figure 4. Suppress of the leaking of ISM components. (**A**) A CP-based SC-ISE by coating the silicon rubber (SR) on the ISM layer. (**B**) A photograph of the SR on the poly(vinyl chloride) (PVC)-based ISM. (**C**) Fourier transform infrared (FTIR) spectra analysis of the SR layer (bottom and top) in comparison with standard ISM components. The ISM components were observed in both bottom and top SR (typical dioctyl sebacate, DOS). (**D**) The K^+-response of SR-coated SC-ISEs compared with an uncoated one. Reprinted with permission from [49], Copyright (2019) American Chemical Society.

3.2. Carbon Materials

Compared with redox CPs, carbon materials emerged for SC materials by virtue of their high conductivity and chemical inertness to environment (gas and light) and redox species. Carbon-based SC materials generally rely on the EDL capacitance response mechanism. In principle, the GC directly coated with ISM like CWEs can be also recognized as the EDL-capacitance ISEs since the GC electrode acts as both solid contact and substrate. An EDL formed between SC and ISM layers. However, the GC electrode shows very small EDL capacitance resulting in the large potential drift according to Equation (6). Therefore, increasing the EDL capacitance is the core subject for the carbon-based SC-ISEs.

In 2007, a 3D-ordered microporous (3DOM) carbon was reported by Stein and Bühlmann groups as the SC material for SC-ISEs [50] (Figure 5A). Owing to the high surface area, the 3DOM carbon-based SC-ISEs showed a low potential drift of 11.7 µV/h. However, the controlled experiments by using highly oriented pyrolytic graphite (HOPG) as the SC disclosed 77 µV/h drift, which demonstrated the high EDL capacitance stabilized mechanism. As expected, there were no gas and light interferences owing to the chemical inertness of carbon materials. After the first introduction of 3DOM carbon as the SC material, they further discussed the effect of architecture and surface chemistry of 3DOM carbon on the ISE performances [51]. The untemplated carbon-based SC-ISEs showed large potential drift further indicating the significance of high EDL capacitance. Meanwhile, the surface oxidized 3DOM carbon with oxygen-containing groups exhibited higher EDL capacitance but the water-layer exists due to the hydrophilic groups. The potential drift was ca. 29 µV/h higher than the unoxidized 3DOM carbon. This work indicates the balance between capacitance and hydrophobicity.

In addition to macroporous carbon, Stein and Bühlmann et al., in 2014 introduced the colloid-imprinted mesoporous (CIM) carbon [52] as the SC material (Figure 5B). The CIM carbon showed nearly 10 times higher capacitance than the 3DOM carbon. The potential drift was further reduced to 1.3 µV/h and the standard derivation of E^0 was only 0.7 mV. Niu's group in 2015 developed porous carbon sub-micrometer spheres (PC-SMSs) [53] for SC-ISEs (Figure 5C). The PC-SMSs featured a high hydrophobicity (contact angle: 137°) and high capacitance (12 mF). The PC-SMSs based K^+-SC-ISEs exhibited 3DOM carbon comparable intermediate-term stability (14.9 µV/h) and excellent long-term stability (ca. 7 mV drift over two months).

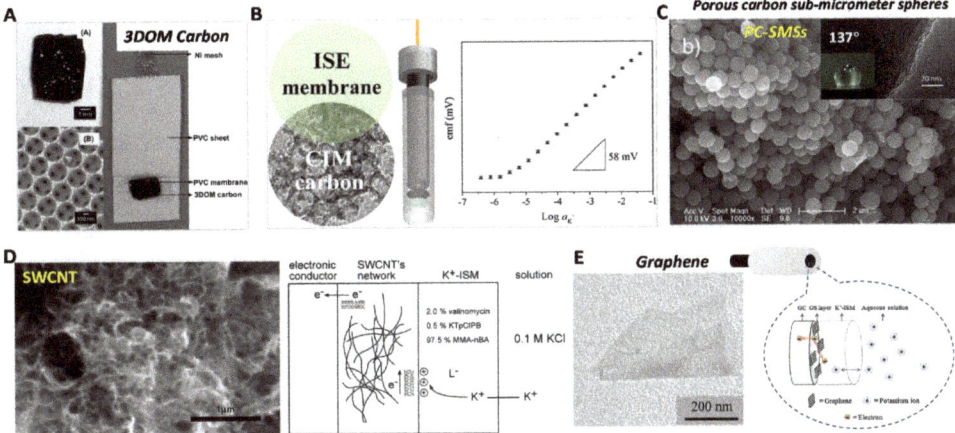

Figure 5. Carbon-based SC-ISEs. (**A**) High-surface 3D-ordered microporous (3DOM) carbon as a SC transducer for SC-ISEs. Reprinted with permission from [50], Copyright (2007) American Chemical Society. (**B**) Colloid-imprinted mesoporous (CIM) carbon with higher surface area as a SC transducer for SC-ISEs. Reprinted with permission from [52], Copyright (2014) American Chemical Society. (**C**) The porous carbon sub-micrometer spheres (PC-SMSs) for superhydrophobic SC transducer with the contact angle up to 137°. Reprinted with permission from [53], Copyright (2015) Elsevier. (**D**) Single-wall carbon nanotube (SWCNT)-based SC-ISEs and the illustrated EDL capacitance response mechanism. Reprinted with permission from [54], Copyright (2008) American Chemical Society. (**E**) The chemically prepared reduced graphene oxide (RGO) as SC transducer for SC-ISEs. Reprinted with permission from [55], Copyright (2012) Royal Society of Chemistry.

Another two representative carbon nanomaterials including carbon nanotubes and graphene have been also explored as the SC materials. In 2008, Rius first introduced single-wall carbon nanotube (SWCNT) [54] to fabricate SC-ISEs (Figure 5D). Nernstian slope and short response time were achieved. Owing to the hydrophobicity of SWCNT, the water-layer was not observed. More importantly, they illustrated in detail that the mechanism can be explained as EDL capacitive coupling [56]. The charged ions are accumulated on the ISM side of the SWCNT/ISM interface, while electrons are located on the SWCNT side, leading to charging/discharging capacitive process. In addition to the cation's sensors, multi-wall carbon nanotube (MWCNT)-based SC-ISEs were reported for anion detection in real environmental samples by the Bakker and Crespo groups [57]. The medium-term stability and interference test were much better than the CPs-based SC-ISEs (like POT).

Niu's group in 2011 introduced graphene sheets [55] as SC materials for K^+-SC-ISEs (Figure 5E). The graphene films were chemically prepared by the reduction of graphene oxide (RGO). Measurements showed an improved water-layer test result and potential stability compared with CWEs. However, the performances particularly for the stability were lower than CNTs and other carbon materials. We attributed this to the existence of hydrophilic surface functional groups (R-COOH, R-OH) on the RGO, leading to a risk of the formation of water layer like the oxidized 3DOM carbon. Wu et al. reported the RGO-based SC-ISEs almost at the same time, which showed the Nernstian response and a low detection of $10^{-6.2}$ M K^+ sensing [58]. Liu et al. modified the graphene oxide by 2-aminowthanethiol and then reduced it to 2-aminoethanethiol functionalized reduced graphene oxide (TRGO) [59]. TRGO could be self-assembled on gold substrate through cohesive Au–S interaction. The cationic and anionic sensing have verified that TRGO-based SC-ISEs have better long-term stability performance under continuous flowing system over two weeks. For the response mechanism, Riu and coworkers demonstrated an EDL capacitive model like other CNTs for graphene-based SC-ISEs [60].

3.3. Other Nanomaterials

In addition to the CPs and carbon materials, a few promising SC materials have been proposed. Niu's group in 2012 developed thiol monolayer-protected Au clusters (MPCs) [61] with tetrakis(4-chlorophenyl) borate (TB$^-$) anion doping for the SC (Figure 6A–C). It should be noted that the Au MPCs is a macromolecule with specific molecular weight. More importantly, the Au MPCs have redox activity and the oxidization and reduced states (MPC$^{+1/0}$) to form the redox buffer. The GC/SC interface is controlled by the ET of MPCs with well-defined phase interfacial potential (Equation (2), Figure 6B). The doped hydrophobic TB$^-$ anion existed in both ISM and SC, performing a reversible IT at the SC/ISM interface. The SC/ISM interfacial potential can be described by Equation (3). The ISM/aq interfacial potential is clear (Equation (4)). Thus, the Au MPCs-based SC-ISEs has clear and well-defined potential definition. The water-layer test demonstrated no water film existed owing to the superhydrophobic thiol protection. The stability showed a deviation of 10.1 ± 0.3 µV/h over continuous monitoring of 72 h (Figure 6C). The standard derivation of E^0 reproductivity was evaluated about ±0.8 mV. This work in one hand proposed redox and superhydrophobic Au MPCs as the SC for SC-ISEs. More importantly, the crucial point is the MPC$^{+1/0}$ redox buffer and TB$^-$ anion doping that offer well-defined phase interfacial potentials.

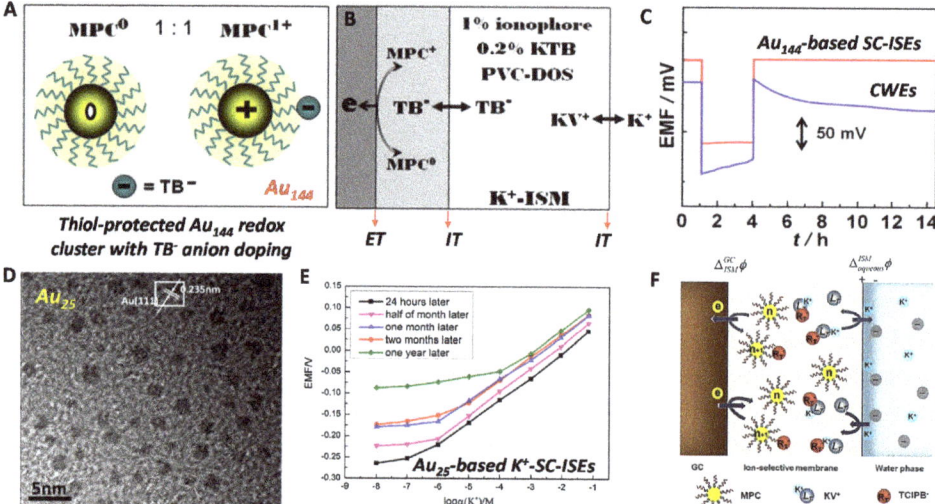

Figure 6. Au nanoclusters-based SC-ISEs. (**A–C**) Au$_{144}$ redox nanocluster based-SC-ISEs. (**A**) Reduced and oxidized states of thiol monolayer-protected Au clusters (MPCs) (Au$_{144}$) doped with tetrakis(4-chlorophenyl) borate (TB$^-$) anion (MPC^0/MPC$^+$TB$^-$). (**B**) Au MPCs-based SC-ISEs with well-defined phase interfacial potential definition. (**C**) Water-layer test of Au$_{144}$-based SC-ISEs. Reprinted with permission from [61], Copyright (2012) American Chemical Society. (**D,E**) Au$_{25}$ redox nanocluster based-SC-ISEs. (**D**) Transmission electronic image (TEM) of the Au$_{25}$ synthesized by optimized one-phase reduction procedure. (**E**) The Au$_{25}$-based SC-ISEs for K$^+$-response to examine its long-term stability. Reprinted with permission from [62], Copyright (2016) Elsevier. (**F**) Au MPCs-based single-piece SC-ISEs through mixing the SC and ISM phase. Reprinted with permission from [63], Copyright (2016) Elsevier.

Regarding to the complex synthesis and low-yield of Au MPCs, Niu's group further proposed one-phase reduction procedure to prepare Au$_{25}$ cluster with high yield [62] (Figure 6D). The Au$_{25}$-based SC-ISEs showed a long-term stability over half a month (Figure 6E). Furthermore, they fabricated a single-piece SC-ISEs by Au MPCs solid contact to simplify the electrode structure [63] (Figure 6F).

A Nernstian response resulted, and the selectivity was not affected. The long-term stability can also last over two weeks.

Like the Au MPCs$^{+1/0}$ redox buffer, Bühlmann et al., in 2013 reported redox Co(II)/Co(III) complex molecules on the thiol-modified Au electrode for the SC-ISEs [64] (Figure 7A). The Co coordination compounds were also doped with TPFPhB$^-$ counter ion to form a well-defined phase boundary potential at the SC/ISM interface determined by TPFPhB$^-$ transfer. The response mechanism is the same with the MPCs-based SC-ISEs. It showed a small standard derivation of E^0 of 1.0 mV over two weeks. Other Co complexes were reported for the SC, which further demonstrated the redox-buffer stabilized concept [65–68]. In addition to Co complexes, tetrathiafulvalene (TTF) and its radical salts have also constituted the redox buffers for SC-ISEs [69]. Schuhmann and coworkers presented an intercalative compound (LiFePO$_4$/FePO$_4$) [70] redox buffer for ion-to-electron transduction (Figure 7B,C). When ion flux comes through the membrane, there is a reversible ion exchange (Fe$^{II/III}$) accompanied by Li$^+$ intercalation/deintercalation. This mechanism provided a well-defined redox interfacial potential and large redox capacitance (Figure 7B). In this configuration, a potential drift of -1.1 ± 1.4 µV/h over 42 days was reported, as well as a standard E^0 deviation of ± 2.0 mV (Figure 7C). This example discovered the growing family of classical lithium battery materials for SC-ISEs.

For the EDL capacitance-type SC-ISE, Qin et al. presented a 3D molybdenum sulfide (MoS$_2$) [71] nanoflower-based SC-ISEs sensor for K$^+$ detections (Figure 7D). The SC material was produced by hydrothermal method, and directly drop-casted on GC electrode. The MoS$_2$-based SC-ISEs showed good Nernstian response and no water layer was observed (Figure 7E). This group further reported an oxygen vacancy MoO$_2$ for SC-ISEs [72]. The oxygen vacancy increased the electronic conductivity and the EDL capacitance, leading to high stability.

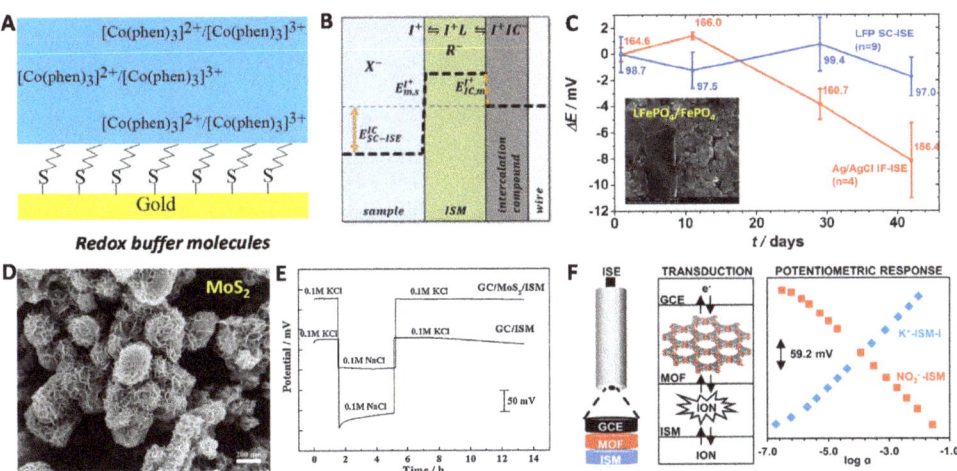

Figure 7. A few other representative SC materials. (**A**) Co(II)/Co(III) complex redox buffer-based SC-ISEs. Reprinted with permission from [64], Copyright (2013) American Chemical Society. (**B,C**) Lithium-battery materials of LiFePO$_4$/FePO$_4$-based SC-ISEs. Reprinted with permission from [70], Copyright (2016) John Wiley and Sons publications. (**D,E**) MoS$_2$ nanomaterials was used for the EDL-type solid contact. SEM image of MoS$_2$ (**D**) and water layer test (**E**). Reprinted with permission from [71], Copyright (2016) Elsevier. (**F**) Metal–organic frameworks (MOFs)-based SC-ISEs. Reprinted with permission from [73], Copyright (2018) American Chemical Society.

In addition to the high surface area carbon materials, metal–organic frameworks (MOFs) [73] with high porous characteristic were first implemented as ion-to-electron transducer by Mirica and coworkers (Figure 7F). The material was synthesized by interconnection of 2,3,6,7,10,11-hexahydroxytriphenylene-HHTP and Ni, Cu, or Co nodes in a Kagome lattice.

The prepared electrode exhibited large bulk capacitance (204 ± 2 µF) and good potential stability (drift of 11.1 ± 0.5 µA/h).

The above MPCs, Co complex and LiFePO$_4$ in principle represent the redox capacitance response mechanism. The MoS$_2$, MoO$_2$ and MOFs can be assigned to the EDL capacitance-based SC-ISEs. Exploring SC transducer materials is still an attractive topic in the field of potentiometric sensing. The realization of SC-ISEs with standard E^0 reproducibility and long-term stability remain urgent requirements.

4. Wearable Sensors

The wearable sensor has become a new concept of ion-selective devices [1,4,6–9,74]. As proposed, biomedical diagnostics are experiencing a change from labor-based implements to portable devices. Wearable ion sensors are one of the portable technologies for healthcare, sports performance monitoring and clinical diagnosis. Compared with traditional LC-ISEs, SC-ISEs reveals the advantages for wearable sensors that can be miniaturized and integrated for continuously detecting the ions in body fluids like sweat, interfacial fluid or saliva in a non-invasive way. Many efforts have contributed to combining SC-ISEs with wearable electronics for real-time bio-fluids monitoring by using garment-, patch-, tattoo-, or sweatband-based objects. In this section, we will introduce recent achievements in the fabrication of SC-ISEs for wearable ion sensors.

4.1. Sweat Ion Detection

Sweat contains abundant chemical information that could evaluate the human body's health state at a biomolecular level according to the medical research. Among which, Na$^+$ and Cl$^-$ are the main ion species ingredients. Beyond that, metabolites, acids, hormones, small proteins and peptides are also the major components of sweat. Their wide fluctuations in an extensive range indicate certain changes in human-body condition. For example, excessive loss of sodium and potassium ions in sweat could result in hyponatremia, hypokalemia, muscle cramps or dehydration. Lithium is an important mood stabilizing drug. They are important biomarkers to evaluate electrolyte balance. Traditional methods for sweat collecting and analyzing separately, would lead to inaccurate analysis due to sample contamination or evaporation in collection process. Real-time collection and detection appear a promising way to achieve accurate measurements.

CPs-based wearable SC-ISEs. Gao and co-workers reported a mechanically flexible fully integrated sensor for simultaneous sensing multiple biomarkers in sweat [75] (Figure 8A). Na$^+$ and K$^+$ sensors were the SC-ISEs with PEDOT(PSS) as the SC, which was synthesized by galvanostatic electrochemical polymerization. The assembled sensor array was fabricated on a flexible polyethylene terephthalate (PET) substrate combined with commercially available integrated-circuit technologies, which allowed for the data visualization in the meantime. This integration technique has established the connection between signal generation, transduction, and visualization without any external multimeter. The chronoamperometric response curves of each component (Na$^+$, K$^+$, glucose, lactate, temperature) were performed separately in different analyte solutions, in which the results showed good potential stability. On-body physical monitoring was also performed. All the indicators exhibited normal characteristics over time, consistent with ex situ measurement results. Besides, the flexible PET substrate meets the requirements of skin friendly. This work opens a breakthrough for the wearable and integrated multiparameter sensors end exhibits the promising real-time on-body monitoring. After that, Javey et al. further extended this concept for the detection of Ca^{2+} and pH in sweat [76] with PEDOT(PSS) as ion-electron transducer (Figure 8B). Ca^{2+} is an important ion related to human metabolism and mineral balance. Ca^{2+} deficiency to a large degree may lead to diseases such as myeloma. They found an interesting result that the Ca^{2+} concentration increases with the decrease of pH in sweat.

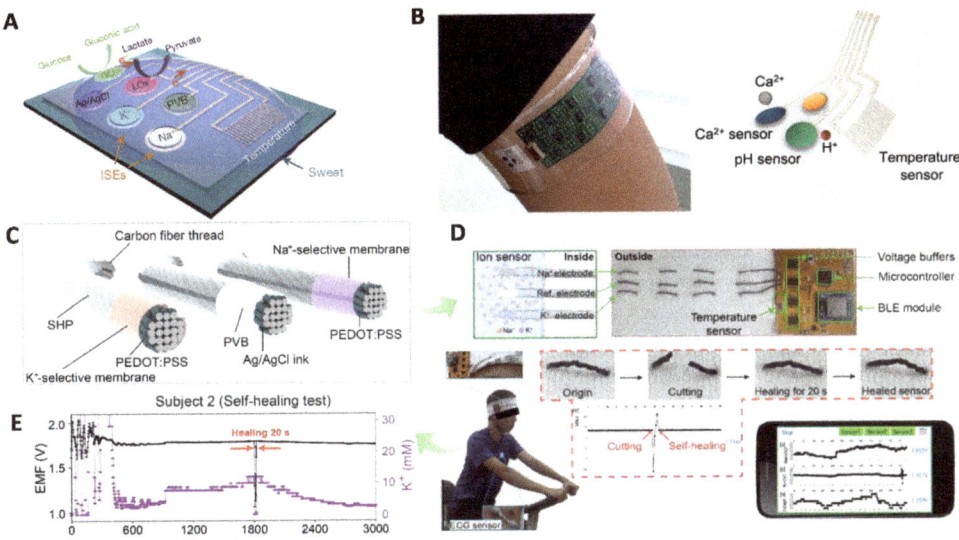

Figure 8. CPs-based wearable SC-ISEs for skin sweat ion sensing. (**A**) An integrated multi-parameter electrochemical wearable sensor involving Na$^+$ and K$^+$-SC-ISEs, glucose, lactate and temperature. The PEDOT was used for the SC transducer. Reprinted with permission from [75], Copyright (2016) Springer Nature. (**B**) PEDOT-based wearable SC-ISEs for pH and Ca^{2+} sensors. Reprinted with permission from [76], Copyright (2016) American Chemical Society. (**C–E**) PEDOT-based self-healable SC-ISEs for wearable senor. Reprinted with permission from [77], Copyright (2019) American Chemical Society.

Yoon et al., very recently proposed a self-healable wearable ion sensor for real-time sweat monitoring [77] (Figure 8C–E). Typically, carbon fiber was used as substrate and PEDOT(PSS) was used as the SC transducer. K$^+$ and Na$^+$ membrane cocktails were dip-coated on the surface of them. On the top of ISM was coated by poly(1,4-cyclohexanedimethanol succinate-*co*-citrate) (PCSC), a supramolecular self-healing polymer (SHP, Figure 8C). The mechanical flexibility of carbon fiber enables the sensor to be attached on many kinds of textile substrates, such as body tattoo, headband, wristband, etc. As shown in Figure 8D, the healed sensor was worn in a tester's head and testing results were real-time checked on a handphone through the integration of a wireless circuit board. The on-body monitoring showed a real-time and accurate reflecting of the user's conditions. When user started cycling, his heart beat fast accompanied with the increase of skin temperature, and then the ion-sensing signal of Na$^+$ began to be observed and increased rapidly and kept steady, while the signal of K$^+$ underwent a gradual increase during cycling. The self-healing ability of the sensor was examined during breaks of the on-body sweat test. After 20 s waiting, the sweat sensor went back to work normally, as before (Figure 8E). This work opens the possibility of a wearable sensor unconstrained by mechanical failure.

Carbon-based wearable SC-ISEs. As discussed in Section 3, in addition to CPs, carbon-based materials are another type of SC-ISEs. As early as 2013, Andrade and coworkers reported a prototype of SWCNT-based wearable SC-ISEs [78] since SWCNT has been demonstrated as an effective EDL-capacitance ion-to-electron transducer [54] in 2008. The authors used the cotton yarns as the flexible substrate filled with SWCNT ink and then coated with ISM (Figure 9A). The sensors showed similar performances to the lab-made SC-ISEs (like the GC substrate). They integrated the sensor on a band-aid and worn on a human model for a proof-of-concept wearable sensing by manually injecting K$^+$ solution. Wang et al. fabricated epidermal tattoo-based wearable SC-ISEs for non-invasive sweat Na$^+$ monitoring by using carbon ink as the SC layer [79]. The sensor can undergo bending and stretching. The sensor was integrated with a wireless signal transduction which was useful for the on-body measurements. In addition to flexible substrates like PET, paper and tattoo, they further reported a textile-based stretchable SC-ISE for both Na$^+$ and K$^+$ sensing by using MWCNT as the SC [80] (Figure 9B). It should be noted that the polyurethane was used instead of PVC for the ISM to improve the mechanical stress and biocompatibility. The sensors could bear 100% strain, 180° bending, crumpling and washing. Zhang et al. reported a carbon textile-based sensor array for real-time sweat monitoring including both ions and another four types of biomarker [81] (glucose, lactate, ascorbic acid and uric acid) (Figure 9C). The N-doped carbon solid-contact was derived from silk fabric. The sensor was integrated with signal collection and transmission via a mobile phone, which realized real-time on-body analysis.

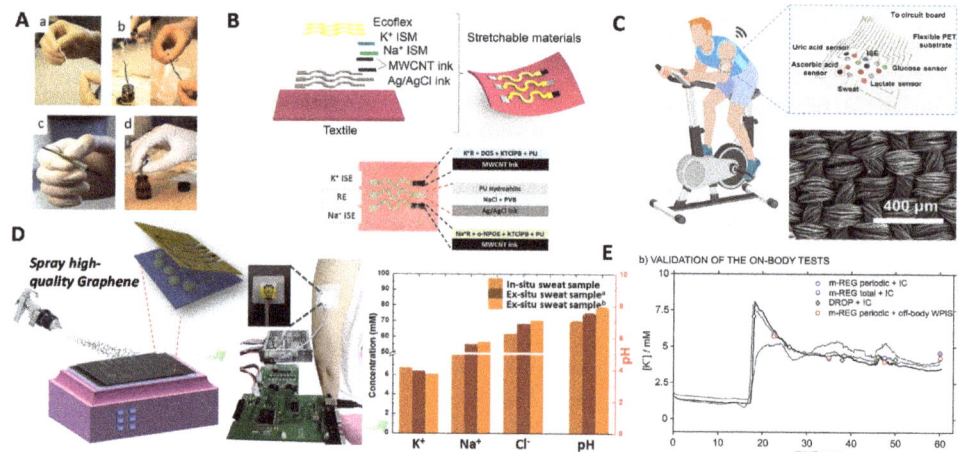

Figure 9. Carbon-based wearable SC-ISEs for skin sweat ion sensing. (**A**) SWCNT-based flexible SC-ISEs by using cotton yard as flexible substrate. Reprinted with permission from [78], Copyright (2013) Royal Society of Chemistry. A series of preparation including SWCNT ink and ISM dipping as shown in (**a–d**). (**B**) Multi-wall carbon nanotube (MWCNT) textile-based SC-ISEs for Na$^+$ and K$^+$ sensing. Reprinted with permission from [80], Copyright (2016) John Wiley and Sons publications. (**C**) Carbon textile-based sensor array for multiparameter analysis. Reprinted with permission from Science Advances [81], Copyright (2019) American Association for the Advancement of Science. (**D**) High-quality graphene-based wearable SC-ISEs for multichannel ion sensing including K$^+$, Na$^+$, Cl$^-$ and pH [82]. (**E**) The suggested protocol for on-body measurement. It should be noted that the importance of calibration of SC-ISEs to assure the accuracy of real-time analysis. Reprinted with permission from [83], Copyright (2019) American Chemical Society.

Niu's group recently reported high-quality graphene-based multichannel wearable SC-ISEs [82] for sweat ion monitoring including Na$^+$, K$^+$, Cl$^-$, and pH (Figure 9D). Compared with reduced graphene oxide (RGO), the high-quality graphene was synthesized by intercalation of graphite. The conductivity and hydrophobicity were much higher than RGO. The flexible paper substrate was superhydrophobic coated by $CF_3(CF_2)_7CH_2SiCl_3$ (C_{10}^F), leading to a water contact angle of 147°. The superhydrophobic substrate effectively prohibited water-layer formation. The flexible sensor showed great structural integrity after being bent for 300 cycles. The real-time on-body monitoring results showed 90% accuracy compared with results verified by ex-situ analysis. Crespo et al. fabricated a four-channel wearable SC-ISEs by using MWCNT as the ion-to-electron transducer [83] (Figure 9E). In this work, importantly, the authors proposed a protocol for the on-body sweat analysis to assure the data validation. They suggested a double validation strategy, i.e., twice off-body measurements before and during on-body analysis. Before on-body measurements, the sweat was measured by wearable SC-ISEs for ex situ analysis. During the on-body measurements, the sweat was sampled at intervals (e.g., every 12.5 min) for comparing analysis by other equipment, for example, ionic chromatography (IC), inductively coupled plasma mass spectrometry (ICP-MS) and pH meter. Moreover, the SC-ISEs should be also calibrated twice before and after on-body measurements. This protocol is highly recommended, which is usefully for evaluating the performance of the wearable SC-ISEs.

Au-based wearable SC-ISEs. With respective to CPs and carbon materials, Au materials disclose more biocompatibility. Zhang et al. recently developed a wearable SC-ISEs sensor based on gold nanodendrite (AuND) array [84] (Figure 10A–C). The AuND array was prepared by electrodeposition on a microwell array patterned chip. Different surface areas of AuND array were fabricated and tested. The results showed that AuND with larger surface area (7.23 cm^2) exhibited enhanced potential stability. Obviously, the stable performance was improved by increasing the surface area of the SC, leading to higher EDL capacitance. The sensors performed an on-body measurement (Figure 10B). The results showed that a steady Na$^+$ response was observed after ca. 15 min warm-up (Figure 10C). During the cycling/rest test, the Na$^+$ level responded periodically.

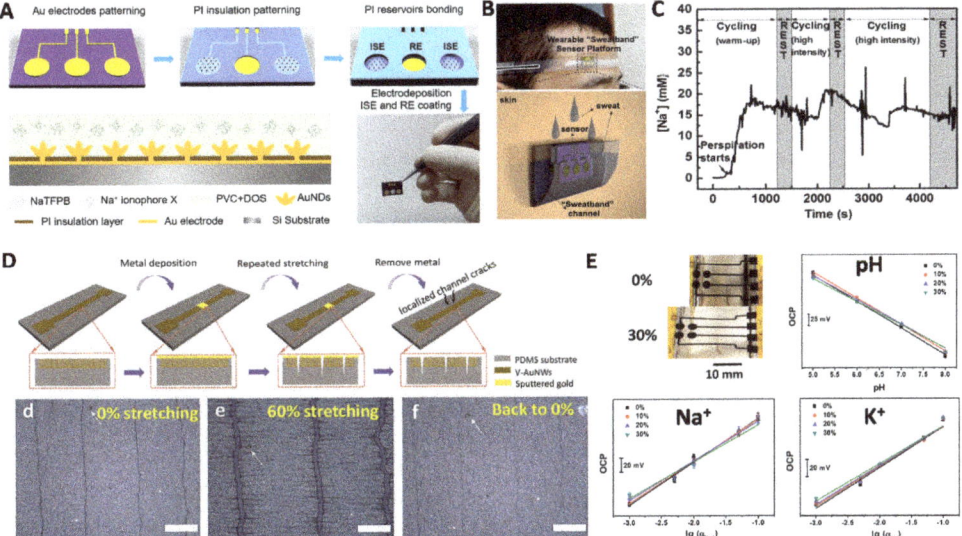

Figure 10. Au nanomaterials-based wearable SC-ISEs for skin sweat ion sensing. (**A–C**) AuND array-based SC-ISEs for sweat Na$^+$ sensing. (**A**) The preparation procedure for AuND-based SC-ISEs by photolithography technique. (**B**) A schematic for the SC-ISEs on-body measurement. (**C**) Real-time analysis of sweat Na$^+$ during cycling and rest states. Reprinted with permission from [84], Copyright (2017) American Chemical Society. (**D,E**) Gold-based vertically aligned nanowires (V-AuNWs)-based stretchable SC-ISEs for Na$^+$, K$^+$ and pH sensing. (**D**) The preparation of V-AuNWs on PDMS film and corresponding optical images for observing the stretching. Scale bar: 200 μm. Reprinted with permission from [85], Copyright (2019) John Wiley and Sons publications. (**E**) The Na$^+$, K$^+$ and pH sensing from 0% to 30% stretching. Reprinted with permission from [86], Copyright (2020) American Chemical Society.

For wearable sensors, the device may often endure with strains. Cheng et al. recently fabricated a stretchable sensor array with gold-based vertically aligned nanowires (V-AuNWs) as the solid contact material for multimodal sweat monitoring including Na$^+$, K$^+$ and pH (Figure 10D,E) [85,86]. The SC-ISEs array was prepared by using PDMS as flexible substrate. The assembled SC-ISEs tattoo was integrated with a flexible printed circuit board (PCB) to achieve wireless real-time monitoring.

The durability of sensors was tested by stretching from 0% to 30% (Figure 10E). The electrode maintained stable potential response.

4.2. Ion Detection in Other Body Fluids

Interstitial fluid is another body fluid that contains a large amount of ingredients related to the human body's health condition. However, the sampling process involves inserting a needle into skin [87] and waiting for the results analyzed by a complex laboratory instrument. This unpleasant process may lead to some complications. It is desirable to develop a device to monitor human physical condition simply and painlessly, especially for patients who take medicine under administration, diabetes patients for example.

Recently, Crespo and co-workers presented a microneedle patch for intradermal detection of potassium in interstitial fluid [88]. The microneedle-based potassium-selective electrode was fabricated with functionalized multiwalled carbon nanotubes (f-MWCNT) and potassium membrane cocktail (Figure 11A–C). These microneedle-based electrodes were fixed in an epidermal patch which was designed to be suitable for insertion into the skin and reached the interstitial fluid in human dermis (Figure 11A). The sensor showed high repeatability in which the potentiometric calibration curves were nearly overlapped before and after 10 insertions (Figure 11B). They performed ex vivo experiments on a piece of chicken skin which was exposed to increasing changes in potassium concentration in an external solution (artificial interstitial fluid) (Figure 11C). As observed, it took ca. 30 min to reach a steady potential response. The difference of concentration between calibration and testing object was attributed to skin state, fat content, individual variation, and the storage condition for the ex vivo test. These factors reveal the challenges for interstitial ion detection.

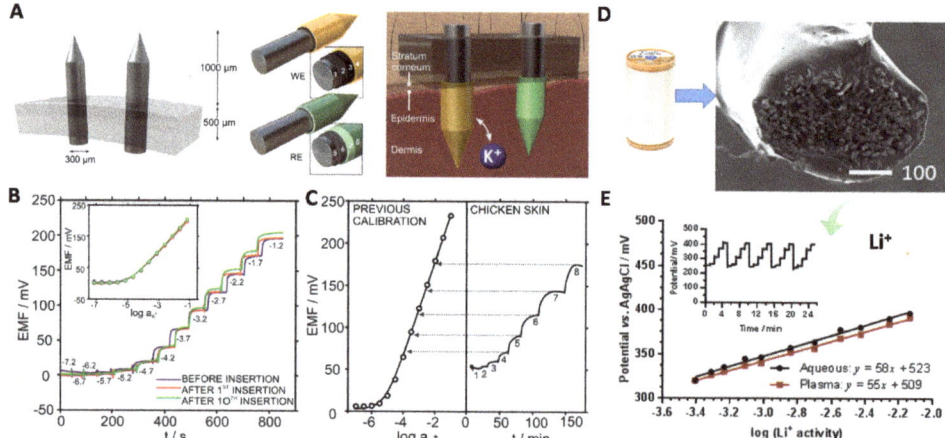

Figure 11. Wearable SC-ISEs for ion detection in interstitial fluid. (**A–C**) Microneedle-based SC-ISEs for K^+ analysis in skin interstitial fluid. (**A**) Illustration of microneedle patch including working electrode (WE) and reference electrode (RE). For the WE, the bare microneedle was coated carbon, f-MWCNTs and K^+-ISM. For the RE, the bare microneedle was coated Ag/AgCl; and poly(vinly butyral) membrane and polyurethane. (**B**) K^+ response before and after insertion into the animal skin. (**C**) Ex vivo K^+ measurement in chicken skin with calibration. Reprinted with permission from [88], Copyright (2019) American Chemical Society. (**D,E**) A cotton fiber-based SC-ISEs for Li^+ sensing in the human plasma. (**D**) A SEM image for the cotton-based SC-ISEs. (**E**) Li^+-response in aqueous solution and human plasma. The inset shows the time traces for Li^+ response. Reprinted with permission from [89], Copyright (2018) American Chemical Society.

Considering the difference of ion concentrations measurements inside and outside of the skin, Crean et al. fabricated a conductive cotton fiber-based SC-ISEs for realizing extraction and analysis simultaneously combined with reverse iontophoresis [89] (Figure 11D,E). This can also be viewed as another kind of in-situ monitoring with simplified approach. The composition of interstitial fluid and human plasma have great in common, except for a higher protein content in the latter. Considering the difficulty of commercial acquisition of interstitial fluid, human plasma was used in this work to test the Li$^+$ response. The electrodes were used in plasma solutions without condition. The EMF response in plasma showed nearly overlapped results compared with aqueous solution (Figure 11E). 2–3 mV decrease compared with that in LiCl solution, because of the interference of Na$^+$. This work opens alternative approach for detecting Li$^+$ drug in human plasma.

Tears, urine and saliva are also the human biological fluids for non-invasive monitoring in SC-ISEs-based wearable sensors. For example, Riu et al. fabricated a disposable SC-ISEs sensor for detecting the K$^+$ in human saliva [90]. The modified hydrophobic SWCNT was used as the transducer. The K$^+$ concentrations of five different saliva samples showed good consistence compared with the results of atomic emission spectrometry measurements. Javey et al. tested Ca^{2+} and pH in urine and tears off-body [76]. The results showed near-Nernstian response. Andrade and co-workers reported an SC-ISEs-based sensor with enhanced recognition property for determining creatinine in diluted urine samples [91]. Overall, these biological fluids also contain abundant health biomarkers, but they might be more suitable for off-body analysis.

5. Conclusions and Outlook

In this review, we have briefly illustrated the development of SC-ISEs including response mechanisms, representative transducer materials and recent achievements in the fabrication of flexible and wearable sensors. There have been undeniable breakthroughs in the field of SC-ISEs aiming at constructing potentiometric wearable sensors with flexibility, mechanical stability, signal stability, and integration, involving novel SC materials and deeper understanding of their response mechanisms. Through persistent exploration of new solid transducer materials, improvements in high signal stability and good reproducibility have been obtained. The detection limit has been lowered and water uptake has been reduced through effective approaches to control the ion-to-electron transducers.

Nowadays, on-body analysis is popular since body fluids provide multiple forms of health-related information. SC-ISEs for wearable sensors monitor an individual's health condition. Remarkable achievements have resulted in improving stability and useful life and flexibility, by incorporating excellent SC transducers with suitable substrates in the form of textiles, tattoos, headbands, etc. The new non-invasive wearable platform shows promising prospects in variety of healthcare-related areas. They provide low-cost, real-time, and precise detection. As we discussed in the final part of the review, reliable analytical method requires synergistic action. For example, flexible substrate materials are expected to adapt user's motion movements without disturbing daily life; SC-ISEs are required to provide accurate and quick monitoring; the integrated circuit board connect with mobile platform reflect ions' concentration or process changes over time. Considering the complex processing of human physiological biomarkers, next generation of sensing platforms are expected to explore new materials and new sensing techniques.

For the outlook of SC-ISEs, we think the following perspective is of significance:

(1) There is an urgent protocol requirement for evaluating the SC-ISEs, for example, a standard measurement protocol for the long-term stability of the potential drift. In addition, evaluation of the redox or EDL-capacitance of the SC materials should provide the specific capacitance for comparison. For the wearable sensors, a standard measure protocol should be also presented as soon as possible including calibration and data validity [83]. (2) In addition to the WE of SC-ISEs, the RE should be given more attention. Currently, polymer-based [92–94], ionic liquid-junction [95–97], and lipophilic salts modified-polymer [98,99] are the main solid-state REs. Recently, a solid-state RE of Ag/AgI based on a self-referencing pulstrode offers a new strategy [100]. (3) The leaking of ISM components is

another issue that should be emphasized. SR coating provides an approach [49]. Development of new ISM membrane or making transducer with ion selectivity are plausible. (4) The biocompatibility of SC transducer and the ISM should be focused upon [101,102]. The ISM component, like the valinomycin for K^+ ionophore, is a neurotoxin. Biocompatibility should be paid sound attention particularly in the process of practical in vivo measurements.

Over the past few decades, we have witnessed the development of ISEs from LC to SC that has miniaturized and integrated the ISEs. In addition to the wearable sensors, SC-ISEs could be further intensively explored for in vivo investigations at cell, tissue or organ level [103–107]. More scientific challenges and the promise of practical application provide tremendous space and the next hot-spot for the SC-ISEs.

Funding: This work was funded by the National Natural Science Foundation of China (21974031, 21974032, 21974033 and 21805052), the Department of Science and Technology of Guangdong Province (2019B010933001) and the Cernet Next Generation Internet Technology Innovation Project (NGII20170113).

Conflicts of Interest: The authors declare no conflict of interest.

Appendix A

For the redox capacitance-based SC-ISEs, there are three interfaces including GC/SC, SC/ISM and ISM/aq (Figure 2A). ET occurs at GC/SC interface with well-defined phase interfacial potential (Equation (2)). For the counter ion transfer at the SC/ISM interface, the interfacial potential (Equation (3)) is derived as follows. The IT of Y^- at SC and ISM are equilibrium,

$$Y^-(SC) \rightleftharpoons Y^-(ISM) \tag{A1}$$

Thus, the Y^- electrochemical potential, $\bar{\mu}_{Y^-}$ in SC and ISM phases is equal, i.e.,

$$\bar{\mu}_{Y^-}(SC) = \bar{\mu}_{Y^-}(ISM) \tag{A2}$$

According to the electrochemical potential expression,

$$\bar{\mu}_{Y^-}(SC) = \mu_{Y^-}^{\ominus\prime}(SC) + RT \ln[Y^-]_{SC} - F\phi(SC) \tag{A3}$$

$$\bar{\mu}_{Y^-}(ISM) = \mu_{Y^-}^{\ominus\prime}(ISM) + RT \ln[Y^-]_{ISM} - F\phi(ISM) \tag{A4}$$

where $\mu_{Y^-}^{\ominus\prime}(SC)$ and $\mu_{Y^-}^{\ominus\prime}(ISM)$ are the conditional chemical potentials in the SC and ISM phases, respectively. $\phi(SC)$ and $\phi(ISM)$ are the phase potentials in SC and ISM phases, respectively.

Combining Equations (A2)–(A4), it can be obtained the SC/ISM interfacial potential E_{ISM}^{SC},

$$E_{ISM}^{SC} = \phi(SC) - \phi(ISM) = \frac{\mu_{Y^-}^{\ominus\prime}(SC) - \mu_{Y^-}^{\ominus\prime}(ISM)}{F} + \frac{RT}{F} \ln \frac{[Y^-]_{SC}}{[Y^-]_{ISM}} \tag{A5}$$

When the $[Y^-]_{ISM}$ and $[Y^-]_{SC}$ equal the unit (like the ET of oxidation and reduction species in the Nernst equation), E_{ISM}^{SC} represents the conditional electrode potential $\Delta_{ISM}^{SC}\phi^{O\prime}(Y^-)$. Thus,

$$E_{ISM}^{SC} = \phi(SC) - \phi(ISM) = \Delta_{ISM}^{SC}\phi^{O\prime}(Y^-) + \frac{RT}{F} \ln \frac{[Y^-]_{SC}}{[Y^-]_{ISM}} \tag{A6}$$

For the K^+ transfer at the ISM/aq interface, similar derivation like Y^- at SC/ISM interface can result in Equation (4) according to the K^+ electrochemical potential equilibrium in ISM and aq phases.

References

1. Kim, J.; Campbell, A.S.; de Ávila, B.E.-F.; Wang, J. Wearable biosensors for healthcare monitoring. *Nat. Biotechnol.* **2019**, *37*, 389–406. [CrossRef] [PubMed]
2. Yang, Y.; Gao, W. Wearable and flexible electronics for continuous molecular monitoring. *Chem. Soc. Rev.* **2019**, *48*, 1465–1491. [CrossRef] [PubMed]
3. Ding, J.W.; Qin, W. Recent advances in potentiometric biosensors. *Trac Trends Anal. Chem.* **2020**, *124*, 115803. [CrossRef]
4. Sempionatto, J.R.; Jeerapan, I.; Krishnan, S.; Wang, J. Wearable Chemical Sensors: Emerging Systems for On-Body Analytical Chemistry. *Anal. Chem.* **2020**, *92*, 378–396. [CrossRef]
5. Shrivastava, S.; Trung, T.Q.; Lee, N.-E. Recent progress, challenges, and prospects of fully integrated mobile and wearable point-of-care testing systems for self-testing. *Chem. Soc. Rev.* **2020**, *49*, 1812–1866. [CrossRef]
6. Bariya, M.; Nyein, H.Y.Y.; Javey, A. Wearable sweat sensors. *Nat. Electron.* **2018**, *1*, 160–171. [CrossRef]
7. Cuartero, M.; Parrilla, M.; Crespo, G.A. Wearable Potentiometric Sensors for Medical Applications. *Sensors* **2019**, *19*, 363. [CrossRef]
8. Legner, C.; Kalwa, U.; Patel, V.; Chesmore, A.; Pandey, S. Sweat sensing in the smart wearables era: Towards integrative, multifunctional and body-compliant perspiration analysis. *Sens. Actuators A Phys.* **2019**, *296*, 200–221. [CrossRef]
9. Parrilla, M.; Cuartero, M.; Crespo, G.A. Wearable potentiometric ion sensors. *Trac Trends Anal. Chem.* **2019**, *110*, 303–320. [CrossRef]
10. Bakker, E.; Bühlmann, P.; Pretsch, E. Carrier-Based Ion-Selective Electrodes and Bulk Optodes. 1. General Characteristics. *Chem. Rev.* **1997**, *97*, 3083–3132. [CrossRef]
11. Cattrall, R.W.; Freiser, H. Coated wire ion-selective electrodes. *Anal. Chem.* **1971**, *43*, 1905–1906. [CrossRef]
12. James, H.J.; Carmack, G.; Freiser, H. Coated wire ion-selective electrodes. *Anal. Chem.* **1972**, *44*, 856–857. [CrossRef] [PubMed]
13. Bobacka, J.; Ivaska, A.; Lewenstam, A. Potentiometric Ion Sensors. *Chem. Rev.* **2008**, *108*, 329–351. [CrossRef] [PubMed]
14. Hu, J.B.; Stein, A.; Buhlmann, P. Rational design of all-solid-state ion-selective electrodes and reference electrodes. *Trac Trends Anal. Chem.* **2016**, *76*, 102–114. [CrossRef]
15. Zdrachek, E.; Bakker, E. Potentiometric Sensing. *Anal. Chem.* **2019**, *91*, 2–26. [CrossRef]
16. Cadogan, A.; Gao, Z.; Lewenstam, A.; Ivaska, A.; Diamond, D. All-solid-state sodium-selective electrode based on a calixarene ionophore in a poly(vinyl chloride) membrane with a polypyrrole solid contact. *Anal. Chem.* **1992**, *64*, 2496–2501. [CrossRef]
17. Veder, J.-P.; De Marco, R.; Patel, K.; Si, P.; Grygolowicz-Pawlak, E.; James, M.; Alam, M.T.; Sohail, M.; Lee, J.; Pretsch, E.; et al. Evidence for a Surface Confined Ion-to-Electron Transduction Reaction in Solid-Contact Ion-Selective Electrodes Based on Poly(3-octylthiophene). *Anal. Chem.* **2013**, *85*, 10495–10502. [CrossRef]
18. Cuartero, M.; Bishop, J.; Walker, R.; Acres, R.G.; Bakker, E.; De Marco, R.; Crespo, G.A. Evidence of double layer/capacitive charging in carbon nanomaterial-based solid contact polymeric ion-selective electrodes. *Chem. Commun.* **2016**, *52*, 9703–9706. [CrossRef]
19. Bobacka, J. Potential stability of all-solid-state ion-selective electrodes using conducting polymers as ion-to-electron transducers. *Anal. Chem.* **1999**, *71*, 4932–4937. [CrossRef]
20. Bobacka, J. Conducting polymer-based solid-state ion-selective electrodes. *Electroanalysis* **2006**, *18*, 7–18. [CrossRef]
21. Bobacka, J.; Ivaska, A.; Lewenstam, A. Potentiometric ion sensors based on conducting polymers. *Electroanalysis* **2003**, *15*, 366–374. [CrossRef]
22. Bobacka, J.; Lindfors, T.; Lewenstam, A.; Ivaska, A. All-solid-state ion sensors, usiong conducting polymers as ion-to-electron transducers. *Am. Lab.* **2004**, *36*, 13.
23. Huang, M.-R.; Gu, G.-L.; Ding, Y.-B.; Fu, X.-T.; Li, R.-G. Advanced Solid-Contact Ion Selective Electrode Based on Electrically Conducting Polymers. *Chin. J. Anal. Chem.* **2012**, *40*, 1454–1460. [CrossRef]

24. Szűcs, J.; Lindfors, T.; Bobacka, J.; Gyurcsányi, R.E. Ion-selective Electrodes with 3D Nanostructured Conducting Polymer Solid Contact. *Electroanalysis* **2016**, *28*, 778–786. [CrossRef]
25. He, N.; Papp, S.; Lindfors, T.; Höfler, L.; Latonen, R.-M.; Gyurcsányi, R.E. Pre-Polarized Hydrophobic Conducting Polymer Solid-Contact Ion-Selective Electrodes with Improved Potential Reproducibility. *Anal. Chem.* **2017**, *89*, 2598–2605. [CrossRef]
26. Papp, S.; Bojtár, M.; Gyurcsányi, R.E.; Lindfors, T. Potential Reproducibility of Potassium-Selective Electrodes Having Perfluorinated Alkanoate Side Chain Functionalized Poly(3,4-ethylenedioxythiophene) as a Hydrophobic Solid Contact. *Anal. Chem.* **2019**, *91*, 9111–9118. [CrossRef]
27. Guzinski, M.; Jarvis, J.M.; D'Orazio, P.; Izadyar, A.; Pendley, B.D.; Lindner, E. Solid-Contact pH Sensor without CO_2 Interference with a Superhydrophobic PEDOT-C_{14} as Solid Contact: The Ultimate "Water Layer" Test. *Anal. Chem.* **2017**, *89*, 8468–8475. [CrossRef]
28. Bobacka, J.; McCarrick, M.; Lewenstam, A.; Ivaska, A. All solid-state poly(vinyl chloride) membrane ion-selective electrodes with poly(3-octylthiophene) solid internal contact. *Analyst* **1994**, *119*, 1985–1991. [CrossRef]
29. Veder, J.P.; Patel, K.; Clarke, G.; Grygolowicz-Pawlak, E.; Silvester, D.S.; De Marco, R.; Pretsch, E.; Bakker, E. Synchrotron Radiation/Fourier Transform-Infrared Microspectroscopy Study of Undesirable Water Inclusions in Solid-Contact Polymeric Ion-Selective Electrodes. *Anal. Chem.* **2010**, *82*, 6203–6207. [CrossRef]
30. Jarvis, J.M.; Guzinski, M.; Pendley, B.D.; Lindner, E. Poly(3-octylthiophene) as solid contact for ion-selective electrodes: Contradictions and possibilities. *J. Solid State Electrochem.* **2016**, *20*, 3033–3041. [CrossRef]
31. Crespo, G.A.; Cuartero, M.; Bakker, E. Thin Layer Ionophore-Based Membrane for Multianalyte Ion Activity Detection. *Anal. Chem.* **2015**, *87*, 7729–7737. [CrossRef] [PubMed]
32. Cuartero, M.; Acres, R.G.; De Marco, R.; Bakker, E.; Crespo, G.A. Electrochemical Ion Transfer with Thin Films of Poly(3-octylthiophene). *Anal. Chem.* **2016**, *88*, 6939–6946. [CrossRef] [PubMed]
33. Cuartero, M.; Crespo, G.A.; Bakker, E. Polyurethane Ionophore-Based Thin Layer Membranes for Voltammetric Ion Activity Sensing. *Anal. Chem.* **2016**, *88*, 5649–5654. [CrossRef] [PubMed]
34. Cuartero, M.; Crespo, G.A.; Bakker, E. Ionophore-Based Voltammetric Ion Activity Sensing with Thin Layer Membranes. *Anal. Chem.* **2016**, *88*, 1654–1660. [CrossRef]
35. Yuan, D.; Cuartero, M.; Crespo, G.A.; Bakker, E. Voltammetric Thin-Layer Ionophore-Based Films: Part 1. Experimental Evidence and Numerical Simulations. *Anal. Chem.* **2017**, *89*, 586–594. [CrossRef] [PubMed]
36. Yuan, D.; Cuartero, M.; Crespo, G.A.; Bakker, E. Voltammetric Thin-Layer Ionophore-Based Films: Part 2. Semi-Empirical Treatment. *Anal. Chem.* **2017**, *89*, 595–602. [CrossRef]
37. Forrest, T.; Zdrachek, E.; Bakker, E. Thin Layer Membrane Systems as Rapid Development Tool for Potentiometric Solid Contact Ion-selective Electrodes. *Electroanalysis* **2020**, *32*, 799–804. [CrossRef]
38. Zdrachek, E.; Bakker, E. Electrochemically Switchable Polymeric Membrane Ion-Selective Electrodes. *Anal. Chem.* **2018**, *90*, 7591–7599. [CrossRef]
39. Liu, C.; Jiang, X.; Zhao, Y.; Jiang, W.; Zhang, Z.; Yu, L. A solid-contact Pb^{2+}-selective electrode based on electrospun polyaniline microfibers film as ion-to-electron transducer. *Electrochim. Acta* **2017**, *231*, 53–60. [CrossRef]
40. Wang, Y.; Xue, C.; Li, X.; Du, X.; Wang, Z.; Ma, G.; Hao, X. Facile Preparation of alpha-Zirconium Phosphate/Polyaniline Hybrid Film for Detecting Potassium Ion in a Wide Linear Range. *Electroanalysis* **2014**, *26*, 416–423. [CrossRef]
41. Huang, Y.; Li, J.; Yin, T.; Jia, J.; Ding, Q.; Zheng, H.; Chen, C.-T.A.; Ye, Y. A novel all-solid-state ammonium electrode with polyaniline and copolymer of aniline/2,5-dimethoxyaniline as transducers. *J. Electroanal. Chem.* **2015**, *741*, 87–92. [CrossRef]
42. Abramova, N.; Moral-Vico, J.; Soley, J.; Ocana, C.; Bratov, A. Solid contact ion sensor with conducting polymer layer copolymerized with the ion-selective membrane for determination of calcium in blood serum. *Anal. Chim. Acta* **2016**, *943*, 50–57. [CrossRef] [PubMed]
43. Boeva, Z.A.; Lindfors, T. Few-layer graphene and polyaniline composite as ion-to-electron transducer in silicone rubber solid-contact ion-selective electrodes. *Sens. Actuators B Chem.* **2016**, *224*, 624–631. [CrossRef]

44. Zeng, X.; Jiang, W.; Jiang, X.; Waterhouse, G.I.N.; Zhang, Z.; Yu, L. Stable Pb^{2+} ion-selective electrodes based on polyaniline-TiO_2 solid contacts. *Anal. Chim. Acta* **2020**, *1094*, 26–33. [CrossRef] [PubMed]
45. He, N.; Hoefler, L.; Latonen, R.-M.; Lindfors, T. Influence of hydrophobization of the polyazulene ion-to-electron transducer on the potential stability of calcium-selective solid-contact electrodes. *Sens. Actuators B Chem.* **2015**, *207*, 918–925. [CrossRef]
46. He, N.; Gyurcsanyi, R.E.; Lindfors, T. Electropolymerized hydrophobic polyazulene as solid-contacts in potassium-selective electrodes. *Analyst* **2016**, *141*, 2990–2997. [CrossRef]
47. Pawlak, M.; Grygolowicz-Pawlak, E.; Bakker, E. Ferrocene Bound Poly(vinyl chloride) as Ion to Electron Transducer in Electrochemical Ion Sensors. *Anal. Chem.* **2010**, *82*, 6887–6894. [CrossRef]
48. Jaworska, E.; Mazur, M.; Maksymiuk, K.; Michalska, A. Fate of Poly(3-octylthiophene) Transducer in Solid Contact Ion-Selective Electrodes. *Anal. Chem.* **2018**, *90*, 2625–2630. [CrossRef]
49. Joon, N.K.; He, N.; Ruzgas, T.; Bobacka, J.; Lisak, G. PVC-Based Ion-Selective Electrodes with a Silicone Rubber Outer Coating with Improved Analytical Performance. *Anal. Chem.* **2019**, *91*, 10524–10531. [CrossRef]
50. Lai, C.-Z.; Fierke, M.A.; Stein, A.; Bühlmann, P. Ion-Selective Electrodes with Three-Dimensionally Ordered Macroporous Carbon as the Solid Contact. *Anal. Chem.* **2007**, *79*, 4621–4626. [CrossRef]
51. Fierke, M.A.; Lai, C.-Z.; Bühlmann, P.; Stein, A. Effects of Architecture and Surface Chemistry of Three-Dimensionally Ordered Macroporous Carbon Solid Contacts on Performance of Ion-Selective Electrodes. *Anal. Chem.* **2010**, *82*, 680–688. [CrossRef]
52. Hu, J.; Zou, X.U.; Stein, A.; Bühlmann, P. Ion-Selective Electrodes with Colloid-Imprinted Mesoporous Carbon as Solid Contact. *Anal. Chem.* **2014**, *86*, 7111–7118. [CrossRef]
53. Ye, J.; Li, F.; Gan, S.; Jiang, Y.; An, Q.; Zhang, Q.; Niu, L. Using sp(2)-C dominant porous carbon sub-micrometer spheres as solid transducers in ion-selective electrodes. *Electrochem. Commun.* **2015**, *50*, 60–63. [CrossRef]
54. Crespo, G.A.; Macho, S.; Rius, F.X. Ion-Selective Electrodes Using Carbon Nanotubes as Ion-to-Electron Transducers. *Anal. Chem.* **2008**, *80*, 1316–1322. [CrossRef] [PubMed]
55. Li, F.; Ye, J.; Zhou, M.; Gan, S.; Zhang, Q.; Han, D.; Niu, L. All-solid-state potassium-selective electrode using graphene as the solid contact. *Analyst* **2012**, *137*, 618–623. [CrossRef] [PubMed]
56. Crespo, G.A.; Macho, S.; Bobacka, J.; Rius, F.X. Transduction Mechanism of Carbon Nanotubes in Solid-Contact Ion-Selective Electrodes. *Anal. Chem.* **2009**, *81*, 676–681. [CrossRef]
57. Yuan, D.; Anthis, A.H.C.; Ghahraman Afshar, M.; Pankratova, N.; Cuartero, M.; Crespo, G.A.; Bakker, E. All-Solid-State Potentiometric Sensors with a Multiwalled Carbon Nanotube Inner Transducing Layer for Anion Detection in Environmental Samples. *Anal. Chem.* **2015**, *87*, 8640–8645. [CrossRef]
58. Ping, J.F.; Wang, Y.X.; Wu, J.; Ying, Y.B. Development of an all-solid-state potassium ion-selective electrode using graphene as the solid-contact transducer. *Electrochem. Commun.* **2011**, *13*, 1529–1532. [CrossRef]
59. Liu, Y.; Liu, Y.; Meng, Z.; Qin, Y.; Jiang, D.; Xi, K.; Wang, P. Thiol-functionalized reduced graphene oxide as self-assembled ion-to-electron transducer for durable solid-contact ion-selective electrodes. *Talanta* **2020**, *208*, 120374. [CrossRef]
60. Hernandez, R.; Riu, J.; Bobacka, J.; Valles, C.; Jimenez, P.; Benito, A.M.; Maser, W.K.; Xavier Rius, F. Reduced Graphene Oxide Films as Solid Transducers in Potentiometric All-Solid-State Ion-Selective Electrodes. *J. Phys. Chem. C* **2012**, *116*, 22570–22578. [CrossRef]
61. Zhou, M.; Gan, S.; Cai, B.; Li, F.; Ma, W.; Han, D.; Niu, L. Effective Solid Contact for Ion-Selective Electrodes: Tetrakis(4-chlorophenyl)borate (TB^-) Anions Doped Nanocluster Films. *Anal. Chem.* **2012**, *84*, 3480–3483. [CrossRef] [PubMed]
62. Xu, J.; Jia, F.; Li, F.; An, Q.; Gan, S.; Zhang, Q.; Ivaska, A.; Niu, L. Simple and Efficient Synthesis of Gold Nanoclusters and Their Performance as Solid Contact of Ion Selective Electrode. *Electrochim. Acta* **2016**, *222*, 1007–1012. [CrossRef]
63. An, Q.; Jiao, L.; Jia, F.; Ye, J.; Li, F.; Gan, S.; Zhang, Q.; Ivaska, A.; Niu, L. Robust single-piece all-solid-state potassium-selective electrode with monolayer-protected Au clusters. *J. Electroanal. Chem.* **2016**, *781*, 272–277. [CrossRef]
64. Zou, X.U.; Cheong, J.H.; Taitt, B.J.; Bühlmann, P. Solid Contact Ion-Selective Electrodes with a Well-Controlled Co(II)/Co(III) Redox Buffer Layer. *Anal. Chem.* **2013**, *85*, 9350–9355. [CrossRef] [PubMed]

65. Zhen, X.V.; Rousseau, C.R.; Bühlmann, P. Redox Buffer Capacity of Ion-Selective Electrode Solid Contacts Doped with Organometallic Complexes. *Anal. Chem.* **2018**, *90*, 11000–11007. [CrossRef] [PubMed]
66. Zou, X.U.; Zhen, X.V.; Cheong, J.H.; Bühlmann, P. Calibration-Free Ionophore-Based Ion-Selective Electrodes With a Co(II)/Co(III) Redox Couple-Based Solid Contact. *Anal. Chem.* **2014**, *86*, 8687–8692. [CrossRef]
67. Jaworska, E.; Naitana, M.L.; Stelmach, E.; Pomarico, G.; Wojciechowski, M.; Bulska, E.; Maksymiuk, K.; Paolesse, R.; Michalska, A. Introducing Cobalt(II) Porphyrin/Cobalt(III) Corrole Containing Transducers for Improved Potential Reproducibility and Performance of All-Solid-State Ion-Selective Electrodes. *Anal. Chem.* **2017**, *89*, 7107–7114. [CrossRef]
68. Jaworska, E.; Pomarico, G.; Berna, B.B.; Maksymiuk, K.; Paolesse, R.; Michalska, A. All-solid-state paper based potentiometric potassium sensors containing cobalt(II) porphyrin/cobalt(III) corrole in the transducer layer. *Sens. Actuators B Chem.* **2018**, *277*, 306–311. [CrossRef]
69. Piek, M.; Piech, R.; Paczosa-Bator, B. Improved Nitrate Sensing Using Solid Contact Ion Selective Electrodes Based on TTF and Its Radical Salt. *J. Electrochem. Soc.* **2015**, *162*, B257–B263. [CrossRef]
70. Ishige, Y.; Klink, S.; Schuhmann, W. Intercalation Compounds as Inner Reference Electrodes for Reproducible and Robust Solid-Contact Ion-Selective Electrodes. *Angew. Chem. Int. Ed.* **2016**, *55*, 4831–4835. [CrossRef]
71. Zeng, X.; Yu, S.; Yuan, Q.; Qin, W. Solid-contact K^+-selective electrode based on three-dimensional molybdenum sulfide nanoflowers as ion-to-electron transducer. *Sens. Actuators B Chem.* **2016**, *234*, 80–83. [CrossRef]
72. Zeng, X.; Qin, W. A solid-contact potassium-selective electrode with MoO_2 microspheres as ion-to-electron transducer. *Anal. Chim. Acta* **2017**, *982*, 72–77. [CrossRef] [PubMed]
73. Mendecki, L.; Mirica, K.A. Conductive Metal-Organic Frameworks as Ion-to-Electron Transducers in Potentiometric Sensors. *Acs Appl. Mater. Interfaces* **2018**, *10*, 19248–19257. [CrossRef]
74. Gao, W.; Ota, H.; Kiriya, D.; Takei, K.; Javey, A. Flexible Electronics toward Wearable Sensing. *Acc. Chem. Res.* **2019**, *52*, 523–533. [CrossRef]
75. Gao, W.; Emaminejad, S.; Nyein, H.Y.Y.; Challa, S.; Chen, K.; Peck, A.; Fahad, H.M.; Ota, H.; Shiraki, H.; Kiriya, D.; et al. Fully integrated wearable sensor arrays for multiplexed in situ perspiration analysis. *Nature* **2016**, *529*, 509–514. [CrossRef] [PubMed]
76. Nyein, H.Y.Y.; Gao, W.; Shahpar, Z.; Emaminejad, S.; Challa, S.; Chen, K.; Fahad, H.M.; Tai, L.-C.; Ota, H.; Davis, R.W.; et al. A Wearable Electrochemical Platform for Noninvasive Simultaneous Monitoring of Ca^{2+} and pH. *Acs Nano* **2016**, *10*, 7216–7224. [CrossRef] [PubMed]
77. Yoon, J.H.; Kim, S.-M.; Eom, Y.; Koo, J.M.; Cho, H.-W.; Lee, T.J.; Lee, K.G.; Park, H.J.; Kim, Y.K.; Yoo, H.-J.; et al. Extremely Fast Self-Healable Bio-Based Supramolecular Polymer for Wearable Real-Time Sweat-Monitoring Sensor. *Acs Appl. Mater. Interfaces* **2019**, *11*, 46165–46175. [CrossRef]
78. Guinovart, T.; Parrilla, M.; Crespo, G.A.; Rius, F.X.; Andrade, F.J. Potentiometric sensors using cotton yarns, carbon nanotubes and polymeric membranes. *Analyst* **2013**, *138*, 5208–5215. [CrossRef]
79. Bandodkar, A.J.; Molinnus, D.; Mirza, O.; Guinovart, T.; Windmiller, J.R.; Valdés-Ramírez, G.; Andrade, F.J.; Schöning, M.J.; Wang, J. Epidermal tattoo potentiometric sodium sensors with wireless signal transduction for continuous non-invasive sweat monitoring. *Biosens. Bioelectron.* **2014**, *54*, 603–609. [CrossRef]
80. Parrilla, M.; Canovas, R.; Jeerapan, I.; Andrade, F.J.; Wang, J. A Textile-Based Stretchable Multi-Ion Potentiometric Sensor. *Adv. Healthc. Mater.* **2016**, *5*, 996–1001. [CrossRef]
81. He, W.; Wang, C.; Wang, H.; Jian, M.; Lu, W.; Liang, X.; Zhang, X.; Yang, F.; Zhang, Y. Integrated textile sensor patch for real-time and multiplex sweat analysis. *Sci. Adv.* **2019**, *5*, eaax0649. [CrossRef] [PubMed]
82. An, Q.; Gan, S.; Xu, J.; Bao, Y.; Wu, T.; Kong, H.; Zhong, L.; Ma, Y.; Song, Z.; Niu, L. A multichannel electrochemical all-solid-state wearable potentiometric sensor for real-time sweat ion monitoring. *Electrochem. Commun.* **2019**, *107*, 106553. [CrossRef]
83. Parrilla, M.; Ortiz-Gómez, I.; Cánovas, R.; Salinas-Castillo, A.; Cuartero, M.; Crespo, G.A. Wearable Potentiometric Ion Patch for On-Body Electrolyte Monitoring in Sweat: Toward a Validation Strategy to Ensure Physiological Relevance. *Anal. Chem.* **2019**, *91*, 8644–8651. [CrossRef] [PubMed]
84. Wang, S.; Wu, Y.; Gu, Y.; Li, T.; Luo, H.; Li, L.-H.; Bai, Y.; Li, L.; Liu, L.; Cao, Y.; et al. Wearable Sweatband Sensor Platform Based on Gold Nanodendrite Array as Efficient Solid Contact of Ion-Selective Electrode. *Anal. Chem.* **2017**, *89*, 10224–10231. [CrossRef] [PubMed]

85. Gong, S.; Yap, L.W.; Zhu, B.; Zhai, Q.; Liu, Y.; Lyu, Q.; Wang, K.; Yang, M.; Ling, Y.; Lai, D.T.H.; et al. Local Crack-Programmed Gold Nanowire Electronic Skin Tattoos for In-Plane Multisensor Integration. *Adv. Mater.* **2019**, *31*, 1903789. [CrossRef]
86. Zhai, Q.; Yap, L.W.; Wang, R.; Gong, S.; Guo, Z.; Liu, Y.; Lyu, Q.; Wang, J.; Simon, G.P.; Cheng, W. Vertically Aligned Gold Nanowires as Stretchable and Wearable Epidermal Ion-Selective Electrode for Noninvasive Multiplexed Sweat Analysis. *Anal. Chem.* **2020**, *92*, 4647–4655. [CrossRef]
87. Romanyuk, A.V.; Zvezdin, V.N.; Samant, P.; Grenader, M.I.; Zemlyanova, M.; Prausnitz, M.R. Collection of Analytes from Microneedle Patches. *Anal. Chem.* **2014**, *86*, 10520–10523. [CrossRef]
88. Parrilla, M.; Cuartero, M.; Padrell Sánchez, S.; Rajabi, M.; Roxhed, N.; Niklaus, F.; Crespo, G.A. Wearable All-Solid-State Potentiometric Microneedle Patch for Intradermal Potassium Detection. *Anal. Chem.* **2019**, *91*, 1578–1586. [CrossRef]
89. Sweilam, M.N.; Varcoe, J.R.; Crean, C. Fabrication and Optimization of Fiber-Based Lithium Sensor: A Step toward Wearable Sensors for Lithium Drug Monitoring in Interstitial Fluid. *Acs Sens.* **2018**, *3*, 1802–1810. [CrossRef]
90. Rius-Ruiz, F.X.; Crespo, G.A.; Bejarano-Nosas, D.; Blondeau, P.; Riu, J.; Rius, F.X. Potentiometric Strip Cell Based on Carbon Nanotubes as Transducer Layer: Toward Low-Cost Decentralized Measurements. *Anal. Chem.* **2011**, *83*, 8810–8815. [CrossRef]
91. Guinovart, T.; Hernández-Alonso, D.; Adriaenssens, L.; Blondeau, P.; Rius, F.X.; Ballester, P.; Andrade, F.J. Characterization of a new ionophore-based ion-selective electrode for the potentiometric determination of creatinine in urine. *Biosens. Bioelectron.* **2017**, *87*, 587–592. [CrossRef] [PubMed]
92. Kisiel, A.; Kijewska, K.; Mazur, M.; Maksymiuk, K.; Michalska, A. Polypyrrole Microcapsules in All-solid-state Reference Electrodes. *Electroanalysis* **2012**, *24*, 165–172. [CrossRef]
93. Zuliani, C.; Matzeu, G.; Diamond, D. A liquid-junction-free reference electrode based on a PEDOT solid-contact and ionogel capping membrane. *Talanta* **2014**, *125*, 58–64. [CrossRef] [PubMed]
94. Lingenfelter, P.; Bartoszewicz, B.; Migdalski, J.; Sokalski, T.; Bucko, M.M.; Filipek, R.; Lewenstam, A. Reference Electrodes with Polymer-Based Membranes-Comprehensive Performance Characteristics. *Membranes* **2019**, *9*, 161. [CrossRef] [PubMed]
95. Cicmil, D.; Anastasova, S.; Kavanagh, A.; Diamond, D.; Mattinen, U.; Bobacka, J.; Lewenstam, A.; Radu, A. Ionic Liquid-Based, Liquid-Junction-Free Reference Electrode. *Electroanalysis* **2011**, *23*, 1881–1890. [CrossRef]
96. Zhang, T.; Lai, C.-Z.; Fierke, M.A.; Stein, A.; Bühlmann, P. Advantages and Limitations of Reference Electrodes with an Ionic Liquid Junction and Three-Dimensionally Ordered Macroporous Carbon as Solid Contact. *Anal. Chem.* **2012**, *84*, 7771–7778. [CrossRef] [PubMed]
97. Zou, X.U.; Chen, L.D.; Lai, C.-Z.; Buehlmann, P. Ionic Liquid Reference Electrodes With a Well-Controlled Co(II)/Co(III) Redox Buffer as Solid Contact. *Electroanalysis* **2015**, *27*, 602–608. [CrossRef]
98. Mattinen, U.; Bobacka, J.; Lewenstam, A. Solid-Contact Reference Electrodes Based on Lipophilic Salts. *Electroanalysis* **2009**, *21*, 1955–1960. [CrossRef]
99. Anderson, E.L.; Chopade, S.A.; Spindler, B.; Stein, A.; Lodge, T.P.; Hillmyer, M.A.; Bühlmann, P. Solid-Contact Ion-Selective and Reference Electrodes Covalently Attached to Functionalized Poly(ethylene terephthalate). *Anal. Chem.* **2020**. [CrossRef]
100. Gao, W.; Zdrachek, E.; Xie, X.; Bakker, E. A Solid-State Reference Electrode Based on a Self-Referencing Pulstrode. *Angew. Chem. Int. Ed.* **2020**, *59*, 2294–2298. [CrossRef]
101. Canovas, R.; Sanchez, S.P.; Parrilla, M.; Cuartero, M.; Crespo, G.A. Cytotoxicity Study of Ionophore-Based Membranes: Toward On Body and in Vivo Ion Sensing. *Acs Sens.* **2019**, *4*, 2524–2535. [CrossRef] [PubMed]
102. Jiang, X.; Wang, P.; Liang, R.; Qin, W. Improving the Biocompatibility of Polymeric Membrane Potentiometric Ion Sensors by Using a Mussel-Inspired Polydopamine Coating. *Anal. Chem.* **2019**, *91*, 6424–6429. [CrossRef] [PubMed]
103. Xu, C.; Wu, F.; Yu, P.; Mao, L. In Vivo Electrochemical Sensors for Neurochemicals: Recent Update. *Acs Sens.* **2019**, *4*, 3102–3118. [CrossRef] [PubMed]
104. Zhao, L.; Jiang, Y.; Wei, H.; Jiang, Y.; Ma, W.; Zheng, W.; Cao, A.-M.; Mao, L. In Vivo Measurement of Calcium Ion with Solid-State Ion-Selective Electrode by Using Shelled Hollow Carbon Nanospheres as a Transducing Layer. *Anal. Chem.* **2019**, *91*, 4421–4428. [CrossRef] [PubMed]
105. Zhao, L.-J.; Zheng, W.; Mao, L.-Q. Recent Advances of Ion-Selective Electrode for in Vivo Analysis in Brain Neurochemistry. *Chin. J. Anal. Chem.* **2019**, *47*, 1480–1491.

106. Zajac, M.; Lewenstam, A.; Bednarczyk, P.; Dolowy, K. Measurement of Multi Ion Transport through Human Bronchial Epithelial Cell Line Provides an Insight into the Mechanism of Defective Water Transport in Cystic Fibrosis. *Membranes* **2020**, *10*, 43. [CrossRef] [PubMed]
107. Zhao, F.; Liu, Y.; Dong, H.; Feng, S.; Shi, G.; Lin, L.; Tian, Y. An Electrochemophysiological Microarray for Real-Time Monitoring and Quantification of Multiple Ions in the Brain of a Freely Moving Rat. *Angew. Chem. Int. Ed.* **2020**, *132*, 10512–10516. [CrossRef]

© 2020 by the authors. Licensee MDPI, Basel, Switzerland. This article is an open access article distributed under the terms and conditions of the Creative Commons Attribution (CC BY) license (http://creativecommons.org/licenses/by/4.0/).

MDPI
St. Alban-Anlage 66
4052 Basel
Switzerland
Tel. +41 61 683 77 34
Fax +41 61 302 89 18
www.mdpi.com

Membranes Editorial Office
E-mail: membranes@mdpi.com
www.mdpi.com/journal/membranes

www.ingramcontent.com/pod-product-compliance
Lightning Source LLC
LaVergne TN
LVHW070614100526
838202LV00012B/643